T0318804

Granular Geomechanics

To Younyee, Jeneva, and Kimberly, with love

Discrete Granular Mechanics Set

coordinated by
Félix Darve

Granular Geomechanics

Matthew R. Kuhn

First published 2017 in Great Britain and the United States by ISTE Press Ltd and Elsevier Ltd

ISTE Press Ltd
27-37 St George's Road
London SW19 4EU
UK

www.iste.co.uk

Elsevier Ltd
The Boulevard, Langford Lane
Kidlington, Oxford, OX5 1GB
UK

www.elsevier.com

Notices

Knowledge and best practice in this field are constantly changing. As new research and experience broaden our understanding, changes in research methods, professional practices, or medical treatment may become necessary.

Practitioners and researchers must always rely on their own experience and knowledge in evaluating and using any information, methods, compounds, or experiments described herein. In using such information or methods they should be mindful of their own safety and the safety of others, including parties for whom they have a professional responsibility.

To the fullest extent of the law, neither the Publisher nor the authors, contributors, or editors, assume any liability for any injury and/or damage to persons or property as a matter of products liability, negligence or otherwise, or from any use or operation of any methods, products, instructions, or ideas contained in the material herein.

For information on all our publications visit our website at http://store.elsevier.com/

British Library Cataloguing-in-Publication Data
A CIP record for this book is available from the British Library
Library of Congress Cataloging in Publication Data
A catalog record for this book is available from the Library of Congress
ISBN 978-1-78548-071-3

Printed and bound in the UK and US

Contents

Foreword

Molecular dynamics is recognized as a powerful method in modern computational physics. This method is essentially based on a factual observation: the apparent strong complexity and extreme variety of natural phenomena are not due to the intrinsic complexity of the element laws but due to the very large number of basic elements in interaction through, in fact, simple laws. This is particularly true for granular materials in which a single intergranular friction coefficient between rigid grains is enough to simulate, at a macroscopic scale, the very intricate behavior of sand with a Mohr–Coulomb plasticity criterion, a dilatant behavior under shearing, non-associate plastic strains, etc. and, in fine, an incrementally nonlinear constitutive relation. Passing in a natural way from the grain scale to the sample scale, the discrete element method (DEM) is precisely able to bridge the gap between micro- and macro-scales in a very realistic way, as it is today verified in many mechanics labs.

Thus, DEM is today in an impetuous development in geomechanics and in the other scientific and technical fields related to grain manipulation. Here lies the basic reason for this new set of books called "Discrete Granular Mechanics", in which not only numerical questions are considered but also experimental, theoretical and analytical aspects in relation to the discrete nature of granular media. Indeed, from an experimental point of view, computational tomography Ð, for example Ð is giving rise today to the description of all the translations and rotations of a few thousand grains inside a given sample and to the identification of the formation of mesostructures such as force chains and force loops. With respect to theoretical aspects, DEM is also confirming, informing or at least precising some theoretical clues such as the questions of failure modes, of the expression of stresses inside a partially saturated medium and of the mechanisms involved in granular avalanches. Effectively, this set has been planned to cover all the experimental, theoretical and numerical approaches related to discrete granular mechanics.

The observations show undoubtedly that granular materials have a double nature, i.e. continuous and discrete. Indeed, roughly speaking, these media respect the matter continuity at a macroscopic scale, whereas they are essentially discrete at the granular microscopic scale. However, it appears that, even at the macroscopic scale, the discrete aspect is still present. An emblematic example is constituted by the question of shear band thickness. In the framework of continuum mechanics, it is well recognized that this thickness can be obtained only by introducing a so-called "internal length" through "enriched" continua. However, this internal length seems to be not intrinsic and to constitute a kind a constitutive relation by itself. Probably, it is because to consider the discrete nature of the medium by a simple scalar is oversimplifying reality. However, in a DEM modeling, this thickness is obtained in a natural way without any ad hoc assumption. Another point, whose proper description was indomitable in a continuum mechanics approach, is the post-failure behavior. The finite element method, which is essentially based on the inversion of a stiffness matrix, which is becoming singular at a failure state, meets some numerical difficulties to go beyond a failure state. Here also, it appears that DEM is able to simulate fragile, ductile, localized or diffuse failure modes in a direct and realistic way Ð even in some extreme cases such as fragmentation rupture.

The main limitation of DEM is probably linked today to the limited number of grains or particles, which can be considered in relation to an acceptable computation time. Thus, the simulation of boundary value problems stays, in fact, bounded by more or less heuristic cases. So, the current computations in labs involve at best a few hundred thousand grains and, for specific problems, a few million. Let us note, however, that the parallelization of DEM codes has given rise to some computations involving 10 billion grains, thus opening widely the field of applications for the future.

In addition, this set of books will also present the recent developments occurring in micromechanics, applied to granular assemblies. The classical schemes consider a representative element volume. These schemes are proposing to go from the macro-strain to the displacement field by a localization operator, then the local intergranular law relates the incremental force field to this incremental displacement field, and eventually a homogenization operator deduces the macro-stress tensor from this force field. The other possibility is to pass from the macro-stress to the macro-strain by considering a reverse path. So, some macroscopic constitutive relations can be established, which properly consider an intergranular incremental law. The greatest advantage of these micromechanical relations is probably to consider only a few material parameters, each one with a clear physical meaning.

This set of around 20 books has been envisaged as an overview toward all the promising future developments mentioned earlier.

Félix DARVE
July 2015

Preface

Granular materials cover much of the earth's surface and are among the most widely used engineering and manufacturing materials. This book focuses on granular soils, in which the dominant particle size is that of sands and gravels. Unlike silts and clays, the behavior of granular soils is almost exclusively the result of short-range, grain-to-grain contact interactions between individual particles. These granular materials exhibit quite unusual characteristics, which make them unique among the many materials that we might encounter, including:

– a strong dependence of strength and stiffness upon the mean confining stress, as imposed by the depth of burial, by the boundary conditions, or by the testing arrangement;

– loading behavior that depends upon the initial fraction of the solid phase – as measured by the void ratio or porosity – and upon the initial arrangement of the particles;

– a tendency for volume expansion (dilatancy) during shear loading;

– a dependence of strength and stiffness on fluid pressure within the void space between solid particles;

– a nearly vanishing elastic domain of deformation, such that most deformation includes (and is often dominated by) a plastic component;

– a bewildering dependence of loading behavior upon the loading history, to the extent that granular materials have defied attempts to achieve a comprehensive constitutive expression;

– constitutive behavior that is scale dependent, such that small and large specimens can exhibit different behaviors.

Each of these unusual aspects is the result of seemingly simple interactions of particle pairs, which collectively result in these fascinating, peculiar and perplexing bulk properties.

The book is concerned with the grain-scale behavior of granular soils, and whenever possible, we apply fundamental discrete mechanics to derive and express this behavior. For this reason, the focus is on discrete rather than continuum mechanics, and although the development and description of constitutive theories are not attempted, bulk quantities such as stress, strain and internal fabric are rigorously cast in their discrete, grain-scale forms. The book opens with a chapter on this topic, which is followed by chapters on grain-to-grain contact interactions, numerical simulation methods, and descriptions of bulk-scale behavior and localization phenomena. As with any book that is limited by publisher constraints and by the writer's own knowledge and capacity, the author regrets not including certain topics that are currently receiving intense interest and study (most notably absent are time-dependent phenomena, multi-phase aspects of behavior, and the effects of particle breakage and cementation). An extensive bibliography is included.

The author would like to thank the editor of this series on granular materials, Félix Darve, who has had the foresight and energy to establish a series of in-depth monographs on this important and rapidly evolving discipline within engineering mechanics.

Matthew R. KUHN

Fabric, Stress, and Strain

1.1. Particle size and shape

The mechanical characteristics of a granular soil depend primarily on volume, size, shape, strength and surface properties of its particles and on the manner in which the particles are arranged. The soil grains have four primary aspects: they have a measurable size; they are relatively rigid objects that interact with neighboring grains through their contacts; their grain movements include rigid rotations; and their interaction forces are limited by friction. These four elements – size, rigidity, rotation and friction – give granular materials their unusual mechanical characteristics. These characteristics include dilatancy (an increase in volume during shear), a response to deviatoric loading that depends on the applied pressure and on the neutral, void-phase mean stress, a thorough dependence of behavior on the loading history, non-associative plastic strain and stress, and a nearly vanishing domain of elastic behavior. Any study of granular materials, therefore, must begin with the grains that comprise the material's solid phase.

The relative volume of the solid phase is usually expressed as the void ratio e, equal to the void volume divided by the volume of the solids. The term "density" connotes the relative magnitude of the void ratio: small values correspond to tightly packed "dense" materials, whereas large values indicate "loose" materials with a more fragile arrangement of particles. Alternative measures of the solid's content are common in other fields: the porosity n is the usual measure in earth sciences (equal to the void volume divided by the total volume), and the solids fraction (one minus the porosity) is used in physical sciences and in material processing industries. Void ratios in the range of 0.4–1.2 are typically used for granular materials. The relative denseness or looseness of a particular arrangement of particles is usually judged in relation to the minimum and maximum void ratios that are achieved by applying standardized preparation procedures to a given soil. The resulting void ratios, e_{min} and e_{max}, depend on size, shape and surface characteristics of the particles. For ideal spherical particles of equal size, the density of a packing can be compared with the densest condition (hexagonal close packing) and with other reference conditions, such as the close random packing, loose random packing and thinnest packing.

Soil particles are rarely spheres of equal or even similar size, and the distribution of sizes is a consequence of the transportation and deposition (or manufacturing) processes that formed the bulk material. Size distribution is measured with sieving and sedimentation methods or by measuring the size of individual particles with various micro-imaging methods. The size of an individual particle, as expressed with a single value, will depend on the measurement technique: with sieving, the opening size; with sedimentation, an equivalent hydraulic radius; and with imaging, a projected dimension or equivalent sphere size. Sieving and sedimentation lead to size distributions that are based on the cumulative weight fractions of material smaller than the given sizes, whereas direct optical measurements yield distributions of the size fraction by the numbers of particles, rather than by their weights. The conventional D_{50} particle size is the size for which 50% of the particles (by their cumulative bulk weight) are smaller. Index C_u, the uniformity coefficient, is the most common measure of size diversity (i.e. lack of uniformity) of particle sizes, and is defined as $C_u = D_{60}/D_{10}$, where sizes D_{60} and D_{10} are indicators of the larger and smaller particle sizes in a bulk material. Both the densest and loosest conditions that can be attained with a given soil, corresponding to e_{min} and e_{max}, respectively, increase with an increasing size diversity (increasing C_u), and the density difference $e_{max} - e_{min}$ also increases with the increasing C_u [YOU 73].

The shape of its particles affect a soil's bulk packing density, strength and stiffness. In natural sands, no two particles have the same shape, and characterizing the shape of even a single particle depends on the scale at which the particle is viewed. At the largest scale, a particle can be judged by its rotundness or *sphericity*, which is quantified by comparing the volumes or radii of the particle's inscribing and circumscribing spheres, with perfect spheres having a value of 1. On the basis of its volume and the volume of the circumscribing sphere, an American football has a sphericity of about 0.4. At a somewhat smaller scale, the *roundness* of a particle is a measure of its pointedness, and is a comparative ratio of the radii of curvature of its gross surface features and of either the particle's mean or inscribed radius [WAD 35, KRU 63, POW 53]. Owing to its less pointed shape, the roundness of a rugby ball is greater, and closer to 1, than an American football. Roundness is qualitatively categorized as well-rounded, rounded, sub-rounded, sub-angular, angular or very angular. Sphericity and roundness are commonly quantified on scales of 0 to 1, based on physical measurement or by comparison using reference silhouettes, such as the chart of Krumbein and Sloss [KRU 63] (Figure 1.1). Cho *et al.* [CHO 06] expressed both sphericity and roundness on a scale of 0 to 1 (with 1 being the most spherical and round) and averaged the two numbers as a measure of the *regularity* of gross particle shape. At a smaller scale, a particle's *texture* is a characterization of its surface roughness. This aspect of shape is the most difficult to measure and express, and it is often modeled by the manner in which it affects the particles' frictional characteristics and their contact stiffnesses (see sections 2.2 and 2.3). Besides sphericity, roundness and texture, Altuhafi *et al.* [ALT 13] have proposed an alternative set of shape descriptors that are suited for an automated analysis of digitized images: sphericity, aspect ratio and convexity. Their sphericity and aspect ratio are closely related to the roundness and sphericities described above (sphericity is derived from the perimeters of particle silhouettes, and aspect ratio is a simple ratio of minimum and maximum projected widths). Their convexity is measured as the ratio of a particle's volume (or its projected area) to the volume of its convex hull. The particles' convexities likely have an important effect on a material's bulk mechanical properties, as a lack of convexity allows multiple contacts between a pair of particles and contributes to their mutual nestling and inter-locking.

Several studies of natural and manufactured sands show that both the minimum and maximum void ratios, e_{min} and e_{max}, decrease with increasing sphericity, roundness and regularity [YOU 73, CHO 06, ROU 08]. The spread of void ratios, $e_{max} - e_{min}$, also increases with these three measures. The effect of particle shape on strength is discussed in section 4.6.2.

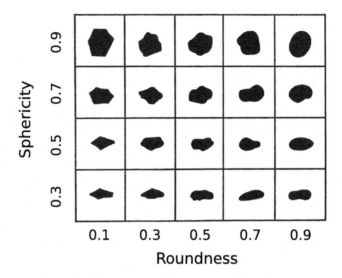

Figure 1.1. *Silhouette chart of Krumbein and Sloss [KRU 63], quantifying roundness and sphericity*

1.2. Granular fabric

Many properties of granular materials are sensitive to the manner in which the particles are arranged, and the *fabric* of a material connotes this internal arrangement of particulate components. A soil's fabric can have a distinct directional character, and as a result, the material can exhibit a marked anisotropy in its mechanical and transport characteristics, as imparted during its formation or during its subsequent loading. The *inherent anisotropy* of fabric and behavior is a consequence of the original manner in which the material was assembled or deposited [ART 77a, ODA 85]. An anisotropy of fabric can impart a mechanical stiffness and strength that depend on loading direction or as hydraulic, electrical or thermal conductivities that depend on the direction of the driving gradient. Although an anisotropic arrangement can be overtly observed as a preferential orientation of the grains and contacts, it can also be subtly present in contact forces and stiffnesses that have a directional bias.

The bulk fabric can be a measurable average of the micro-scale particle arrangements (the subject of this section), or it can be a conceptual phenomenological quantity, often a tensor, that is attributed to the granular arrangement and imparts an anisotropic character to the continuum constitutive description of a material (e.g. the fabric tensor of Li and Dafalias [LI 02b]). Distinctions are also made between the terms fabric and structure, as the latter encompasses fabric but also includes the manner in which internal stress is borne by the forces between the particles. This force-transmission aspect of a granular material is the subject of section 1.3.

The mechanical effect of a soil's initial fabric was clearly demonstrated in experiments by Masanobu Oda, undertaken during the 1970s [ODA 72a]. In one experiment, three sand specimens were prepared with different techniques, even though some specimens shared the same initial void ratio. His plots of stress, strain and volume change during triaxial compression for the three specimens were quite different, even for those specimens having the same initial density. That is, different preparation techniques produced different internal fabrics, resulting in different mechanical behaviors. In other experiments, similar to those of Arthur and Menzies [ART 72], a cylindrical mold was submerged at different orientation angles φ, and sand grains were poured vertically to create samples with different preferred orientations of the grains relative to the cylinder's axis [ODA 72a]. The responses to triaxial compression loading along the cylinder axis displayed markedly different stiffnesses and strengths, which were attributed to the specimens' initial anisotropic fabrics.

1.2.1. *Spatial description*

Although a granular medium is unequivocally partitioned into solid and void regions, an entire region can also be partitioned (tessellated) in other ways. Specifically, a region can be partitioned into a covering of sub-regions, with each sub-region containing both solid and void. In section 1.2.6, a topologic approach is applied to the partitioning of granular regions, with particular attention given to arrangements of the contacts among particles. In this section, we consider only the geometric, spatial arrangement of the particles. In this regard, the simplest partition is the Voronoi tessellation of a set of discrete points into a spatial covering of polygons (in two dimensions) or polyhedra (in three dimensions), as shown in Figure 1.2a. Each sub-region is associated with a single particle, in particular the reference point that has been assigned to the particle, such that all points within the sub-region are nearer to this point than to the reference points of other sub-regions. This tessellation is particularly suited to assemblies of mono-disperse (i.e. single size) disks or spheres in which the reference points are at the particles' centers, since each sub-region fully encompasses a single particle. The edges (two dimensions) or faces (three dimensions) are the perpendicular bisectors of the reference points of neighboring particle pairs. The Voronoi tessellation results in convex sub-regions that have the important *additive* property: the sub-region that is nearest to a *set* of particles is the union of the individual Voronoi sub-regions of these particles. With mono-disperse assemblies of disks (or spheres), every contact between a pair of particles is associated with a single edge (face) of the neighboring pair of polygons (polyhedra), although other edges (faces) might not correspond to contacts and can be considered as "virtual contacts" of nearby, but non-contacting disks (spheres). The stress associated with a single polygon or polyhedron can be readily computed from the contact forces (section 1.3.1). By connecting all pairs of reference points that share common edges (two dimensions) or common faces (three dimensions), we identify the edges of a system of simplices (triangles or tetrahedra) that also cover the same region (Figure 1.2(c)). This Delaunay tessellation is the topologic dual of the Voronoi tessellation and can be used for computing local strains within a granular material (section 1.4.1.1).

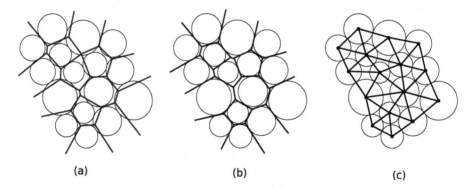

(a)   (b)   (c)

Figure 1.2. *Spatial partitions of a granular region: (a) Voronoi tessellation with respect to the centers of particles; (b) radical Voronoi tessellation of disks; (c) dual, Delaunay tessellation*

The Voronoi partition is less suited, however, for assemblies of poly-disperse (diversely sized) disks or spheres, since the sub-region of a large particle might not fully contain the particle (as in Figure 1.2a). The radical Voronoi tessellation is more appropriate for poly-disperse disks and spheres, as each sub-region encompasses a single particle (Figure 1.2b). The sub-regions are separated by radical planes that are associated with two nearby particles, such that two lines of equal length join any point on the plane with the lines' tangent points on the two particles. The radical Voronoi partition can also be applied to particles whose shapes are approximated to clusters of disks (or spheres), as shown in Figure 3.1f, by finding the union of the radical Voronoi polygons (polyhedra) associated with each component of the cluster particle.

Efficient algorithms have been developed for partitioning space with the Voronoi tessellation of a set of discrete points (particle centers) and with the radical Voronoi tessellation of a set of points and their associated radial sizes. With non-disk and non-sphere particles, efficiently partitioning space into sub-regions that are associated with the individual particle bodies becomes much more difficult. In principle, we can identify a sub-region that is nearer to a single particle than to all other particles, although algorithms have not yet been developed for general particle shapes. Stereological methods have been applied to this purpose, when the region and its particles are expressed with digitized images.

1.2.2. *Topological description*

A granular material can be viewed as a collection (set) of particles that are associated with each other through their contacts (and, perhaps, through the virtual contacts between nearby but non-touching particles), and the topology of a granular assembly is defined by this set of particles and their pairwise associations. The topology is succinctly described with a *particle graph* of the particles (represented by graph vertices) and contacts (graph edges), as in Figure 1.3a. The graph of a two-dimensional material is planar, and its dual graph is the *void graph* in which the vertices are voids and edges are contacts (Figurs 1.3b). Although the graph of a three-dimensional material is not planar and lacks a distinct dual, the graph is enriched with topological objects that can be identified as particles, contacts, pore bodies and pore throats (section 1.2.6.2).

Using computational geometry techniques, the particle graph of a granular assembly is expressed with an adjacency matrix $[\mathbf{A}^{adj}]$. The matrix is symmetric and square, with one row and one column for each particle (vertex). The matrix is filled with 0's and 1's: a 1 in position A_{pq}^{adj} designates an edge (contact) between particles p and q, whereas a 0 means that the particles are not in contact. Weighted graphs, in which numerical weights are assigned to edges in place of the 1's, are applied in other sections and they are used in many disciplines to analyze nearest-distance and least-time routing problems. Besides providing the full topologic description of a granular medium, several scalar descriptors are readily extracted from $[\mathbf{A}^{\text{adj}}]$. The number of particles that touch a particle p (i.e. the *coordination number* of p) is simply the sum of numbers in column (or row) p of $[\mathbf{A}^{\text{adj}}]$. The total number of touching pairs is half the sum of all numbers in the matrix.

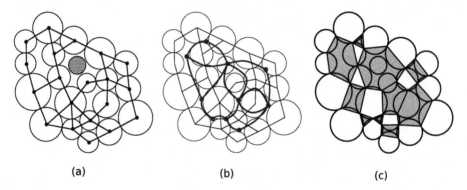

(a)   (b)   (c)

Figure 1.3. *Planar graphs of two-dimensional region: (a) particle graph, in which the contacting particles are the graph's nodes (note the shaded rattler particle); (b) void graph, in which the nodes represent void polygons and (c) K–N partition of void and solid space (section 1.2.6.1)*

1.2.3. *Directional description*

Describing a granular material's fabric involves characterizing four different geometric components of the material: the particle bodies, the surfaces of the particles, the contacts between particles and the void space among the particles. The characteristics of these components encompass both directional and non-directional (scalar) qualities. Although particle shape was discussed in the previous section, in this section, we describe general directional measures of fabric and its anisotropy, and these measures are then applied to bodies, surfaces, contacts and voids in later sections. Bulk, macro-scale averages of these characteristics can be measured for an entire material region (the primary focus of this section), but fabric can vary locally, at both the micro- (particle) and meso- (particle cluster) scales. A full accounting of fabric should also account for its diversity and inhomogeneity (Chapter 4).

The directional nature of a material's fabric can be expressed in a number of ways: for example, with histograms or with fabric tensors. The basis of these measures is the probability of encountering a direction $\mathbf{n}$ and the associated probability density $f(\mathbf{n})$ of this direction. If characteristic g also depends on the orientation and has the mean value $g(\mathbf{n})$ for orientation $\mathbf{n}$, then its average value among all orientations, designated as $\bar{g}$ or $\langle g(\mathbf{n}) \rangle$, is the probability-weighted average:

$$\bar{g} \text{ or } \langle g(\mathbf{n}) \rangle = \int_{\Omega} g(\mathbf{n}) f(\mathbf{n}) \, d\Omega \qquad [1.1]$$

The density $f(\mathbf{n})$ is normalized, so that its own weighted average $\langle 1 \rangle = 1$. Integration is over the unit sphere Ω of orientations $\mathbf{n}$ (unit circle in two dimensions). Ken-Ichi Kanatani [KAN 84] approximated the distribution $f(\mathbf{n})$ as the sum of even-rank tensors $\mathbf{D}^{\bar{m}}$ (of rank m) multiplied by the Cartesian components n_i of direction vector $\mathbf{n}$:

$$f(\mathbf{n}) \approx \frac{1}{Q} \left(1 + D_{ij}^{\bar{2}} n_i n_j + D_{ijkl}^{\bar{4}} n_i n_j n_k n_l + \dots \right) \qquad [1.2]$$

where Q is 2π for two-dimensional systems and 4π in three-dimensional systems, and the overline numerals are the tensors' ranks, not exponents. Kanatani referred to $\mathbf{D}^{\bar{m}}$ as a *fabric tensor of the third kind* of rank m. Fidelity to the true distribution is improved, of course, by extending the series in equation [1.2] with additional terms. The approximation is composed solely of even-rank products of the components n_i, implying a central symmetry and no skew symmetry, $f(\mathbf{n}) = f(-\mathbf{n})$. All of the $\mathbf{D}^{\bar{m}}$ tensors are deviatoric and account for anisotropy in the distribution $f(\mathbf{n})$, whereas the leading term $1 \times 1 / Q$ is the sole isotropic contribution to the density. Kanatani [KAN 84] derived explicit forms of the rank 2, 4 and 6 tensors $\mathbf{D}^{\bar{m}}$, which are computed from certain data-tensors $\mathbf{N}^{\tilde{m}}$. For three-dimensional systems,

$$D_{ij}^{\tilde{2},\,3D} = \frac{15}{2}\left[N_{ij}^{\tilde{2}} - \frac{1}{3}\delta_{ij}\right] \tag{1.3}$$

$$D_{ijkl}^{\tilde{4},\,3D} = \frac{315}{8}\left[N_{ijkl}^{\tilde{4}} - \delta_{ij}N_{kl}^{\tilde{2}} + \frac{1}{8}\delta_{ij}\delta_{kl}\right] \tag{1.4}$$

$$D_{ijklpq}^{\tilde{6},\,3D} = \frac{3003}{16}\left[N_{ijklpq}^{\tilde{6}} - \frac{3}{2}\delta_{ij}N_{klpq}^{\tilde{4}} + \frac{9}{16}\delta_{ij}\delta_{kl}N_{pq}^{\tilde{2}} - \frac{5}{231}\delta_{ij}\delta_{kl}\delta_{pq}\right] \tag{1.5}$$

and for two-dimensional systems,

$$D_{ij}^{\tilde{2},\,2D} = 4\left[N_{ij}^{\tilde{2}} - \frac{1}{2}\delta_{ij}\right] \tag{1.6}$$

$$D_{ijkl}^{\tilde{4},\,2D} = 16\left[N_{ijkl}^{\tilde{4}} - \delta_{ij}N_{kl}^{\tilde{2}} + \frac{1}{8}\delta_{ij}\delta_{kl}\right] \tag{1.7}$$

$$D_{ijklpq}^{\tilde{6},\,2D} = 64\left[N_{ijklpq}^{\tilde{6}} - \frac{3}{2}\delta_{ij}N_{klpq}^{\tilde{4}} + \frac{9}{16}\delta_{ij}\delta_{kl}N_{pq}^{\tilde{2}} - \frac{1}{32}\delta_{ij}\delta_{kl}\delta_{pq}\right] \tag{1.8}$$

The $\mathbf{N}^{\tilde{m}}$ tensors are computed as the averaged products of the n_i components among the M data values:

$$N_{i_1 i_2 i_3 \ldots i_{\tilde{m}}}^{\tilde{m}} = \frac{1}{M}\sum_{\ell=1}^{M} n_{i_1}^{\ell} n_{i_2}^{\ell} n_{i_3}^{\ell} \cdots n_{i_{\tilde{m}}}^{\ell} \tag{1.9}$$

For two-dimensional systems, the distribution of equation [1.2] is expressed more directly as a Fourier series in the sines and cosines of angle θ measured from the n_1 direction, with $[n_1, n_2] = [\cos(\theta), \sin(\theta)]$. The series is

$$f(\mathbf{n}) \approx \frac{1}{2\pi}\,[1 + a_2\cos(2\theta) + b_2\sin(2\theta)$$

$$+\, a_4\cos(4\theta) + b_4\sin(4\theta) + \ldots\,] \tag{1.10}$$

with coefficients $a_2 = D_{11}^{\tilde{2}}$, $b_2 = D_{12}^{\tilde{2}}$, $a_4 = D_{1111}^{\tilde{4}}$, $b_4 = D_{1112}^{\tilde{4}}$, $a_6 = D_{111111}^{\tilde{6}}$ and $b_6 = D_{111112}^{\tilde{6}}$ [KAN 84].

1.2.4. *Particle-body and particle-surface orientations*

In the early 1970s, decades before modern tomographic methods allowed for the non-destructive visualization of soil's internal structure, Oda conducted a series of remarkable experiments on sand specimens, in which the particles were "frozen" in place with a hardening resin and then microscopically viewed as thin sections [ODA 72a, ODA 72b]. The experiments revealed the directional character of the particles' orientations imparted by different deposition methods. Oda also showed that the initial distribution of particle orientations was altered by subsequent loading and was different inside and outside of the shear bands that had formed during loading.

The orientation of a particle can be defined with a triad of orthogonal directions. These directions are clearly identified for a symmetrically shaped particle (e.g. an ellipsoid); otherwise, the directions can be taken as the particle's principal axes of inertia. For each particle p, the widths in these directions are designated as $[a_1^p, a_2^p, a_3^p]$, in the order of decreasing width, and the corresponding unit directions are $[\mathbf{q}_1^p, \mathbf{q}_2^p, \mathbf{q}_3^p]$. We can then associate two matrices with each particle p:

$$\mathbf{A}^p = \begin{bmatrix} a_1^p & 0 & 0 \\ 0 & a_2^p & 0 \\ 0 & 0 & a_3^p \end{bmatrix}, \quad \mathbf{J}^p = \begin{bmatrix} \mathbf{q}_1^p, \ \mathbf{q}_2^p, \ \mathbf{q}_3^p \end{bmatrix} \qquad [1.11]$$

with $\mathbf{A}^p$ providing the widths, and the columns of $\mathbf{J}^p$ providing the corresponding orientation vectors[1]. When averaged among all N particles in an assembly, these quantities define a tensor-valued measure of the average particle orientation:

$$\bar{J}_{ij} = \frac{1}{N} \sum_{p=1}^{N} \frac{3}{\mathrm{tr}\left(A_{ij}^p\right)} J_{ik}^p A_{kl}^p J_{jl}^p \qquad [1.12]$$

where $\mathrm{tr}(A_{ij}^p)$ is the trace of $\mathbf{A}^p$. Tensor $\bar{\mathbf{J}}$ is similar to the orientation tensor of Oda [ODA 85], but equation [1.12] includes the factors $3/\mathrm{tr}(\mathbf{A}^p)$ that eliminate the bias of particle size, so that each particle is given an equal

---

1 Note that the orientation of a particle can also be specified with a quaternion, as described in section 3.1.6

weight, regardless of its size. The factor $1/N$ normalizes the tensor, so that an assembly of spheres would yield the identity (Kronecker) tensor for $\overline{\mathbf{J}}$. When the orientation of a particle can be characterized with a single direction $\mathbf{q}_1^p$ (as with solids of revolution), the distribution density of these directions within an assembly can be represented with fabric tensors $\mathbf{D}^{\bar{m}}$, as described in section 1.2.3. Ouadfel and Rothenburg [OUA 01] expressed the distribution of the elongated directions l_i^p of prolate spheroids in the form of equation [1.3], and Rothenburg and Bathurst [ROT 93] represented directions of the longer axis of two-dimensional ellipses in the form of equation [1.10], with an elongation parameter a_e in place of a_2.

Kuo et al. [KUO 98] introduced a similar measure of anisotropy based on the orientations of the surfaces of the particles. This measure is expressed in the form of equation [1.2] by approximating the total projected surface area (per unit of volume) of the particles in direction n_i,

$$S(\mathbf{n}) \approx \frac{S_v}{4\pi V}\left(1 + \overline{Q}_{ij}n_i n_j\right) = \frac{S_v}{4\pi V}\left(1 + \mathbf{n} \cdot (\overline{\mathbf{Q}} \cdot \mathbf{n})\right) \qquad [1.13]$$

where $S(\mathbf{n})$ is a distribution function, $\overline{\mathbf{Q}}$ is a surface area (fabric) tensor and S_v is the total surface area of particles in volume V. They used stereological methods to estimate $S(\mathbf{n})$ from two-dimensional images along three orthogonal planes. When full descriptions of the particles are available, as in DEM simulations, a similar average orientation tensor $\overline{\mathbf{S}}$ can be directly computed by integrating the dyads $n_i n_j$ across the surface ∂S^p of each particle p for all of the N particles in an assembly:

$$\overline{S}_{ij} = \frac{3}{S_N}\sum_{p=1}^{N}\int_{\partial\Omega^p} n_i n_j \, dS^p \quad \text{or} \quad \overline{\mathbf{S}} = \frac{3}{S_N}\sum_{p=1}^{N}\int_{\partial\Omega^p} \mathbf{n} \otimes \mathbf{n} \, dS^p \qquad [1.14]$$

in which $\partial\Omega^p$ is the surface of particle p with incremental area dS^p and S_N is the total area $\sum\int dS$ of the particles. In this definition, the tensor is normalized so that $\overline{\mathbf{S}}$ is the identity matrix for an assembly of spheres. If we divide by the total mass instead of by S_N, equation [1.14] yields a corresponding tensor measure of specific surface (surface area per unit of volume) that incorporates a material's directional character.

1.2.5. *Contact fabric*

Although the orientations of its particles certainly affect (and are altered by) the loading behavior of a granular material, several researchers in the 1970s and the early 1980s recognized that particle interactions, which take place through the contacts between particles, are strongly influenced by the directional character of the contacts [ODA 72a, KAN 79, CHR 81, SAT 82]. This insight led to the characterization of granular materials by the number and orientation of their contacts, the latter with so-called fabric tensors of various forms. The number of contacts is usually expressed with the average coordination number $\bar{n}$, which is the average number of contacts per particle and is equal to twice the number of contacts M divided by the number of particles N:

$$\bar{n} = 2\frac{M}{N} \qquad [1.15]$$

A large coordination number generally indicates more stable and structurally redundant arrangements of particles (section 4.3), and Oda [ODA 77] showed that the initial average coordination number is positively correlated with the peak deviatoric strength. Oda [ODA 77] also proposed that the standard deviation of the coordination numbers of individual particles is a measure of the heterogeneity of the particle arrangement. Cresswell and Powrie [CRE 04] noted that natural sands with irregular particle shapes tend to produce an abundance of both nestled and flat contacts in which particles touch across relatively large areas of their surfaces, rather than the point-like contacts between convex surfaces that are idealized in most numerical simulations. To characterize the surface areas of the nestled and flat contacts, Fonseca *et al.* [FON 13] proposed a contact index (CI) that expresses the fraction of contacting surface area relative to the full particle surface area, and they found that some sands of geologic origin had an average of 30% of their area occupied in contact.

Figure 1.4 shows two particles, p and q, that share two contacts. A particle is assigned a reference material point, χ^p for particle p, that is attached to the particle, usually (but not necessarily) at its center. The orientation of a single contact c is given by the unit normal vector $\mathbf{n}^c$ that is perpendicular to the contact's tangent plane and points outward from the designated particle of the pair (particle p of the ordered pair p–q). The most common bulk measure of the orientations of all contacts within an assembly is the Satake fabric tensor, which is formed from the first two terms of equation [1.2], when applied to the directions $\mathbf{n}^c$ of the contacts:

$$\overline{F}_{ij} = \frac{1}{M} \sum_{c=1}^{M} n_i^c n_j^c \qquad [1.16]$$

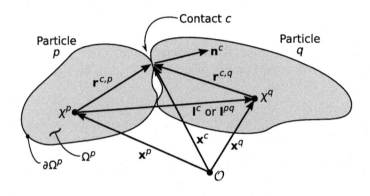

Figure 1.4. *Contact quantities associated with a contact c between two particles, p and q: the contact normal $\mathbf{n}^c$, the contact branch vector $\mathbf{l}^c$ and the contact vectors $\mathbf{r}^{c,p}$ and $\mathbf{r}^{c,q}$*

The sum is averaged over the M contacts of a granular assembly, and the overline designates a bulk, averaged quantity. The tensor $\overline{\mathbf{F}}$ is of type $\mathbf{N}^2$ in equations [1.3] and [1.9], and with equation [1.2], the probability density $f^c(\mathbf{n})$ of contacts having orientation $\mathbf{n}$ is approximated as:

$$f^c(\mathbf{n}) \approx \frac{15}{8\pi}\left(\overline{F}_{ij}n_i n_j - \frac{1}{5}\right) \quad \text{and} \quad \frac{2}{\pi}\left(\overline{F}_{ij}n_i n_j - \frac{1}{4}\right) \qquad [1.17]$$

in three and two dimensions, respectively. Although the Satake tensor is the most commonly used of all fabric tensors, alternative orientation tensors are derived from other vector quantities associated with the contacts. In particular, we can use the branch vector $\mathbf{l}^c$ that joins the particles' two reference points (χ^p and χ^q) or the contact vectors $\mathbf{r}^{c,p}$ and $\mathbf{r}^{c,q}$ that join the reference points of

p and q with the contact c (Figure 1.4). When expressed as a unit vector, each type of vector is associated with a fabric tensor, for example,

$$\overline{G}_{ij} = \frac{1}{M} \sum_{c=1}^{M} (l_i^c l_j^c)/|\mathbf{l}^c|^2 \tag{1.18}$$

which has been suggested by Magoariec *et al.* [MAG 08] as a possible internal variable for predicting the stress in two-dimensional assemblies of ellipses. This measure also admits possible correlations between the orientations and magnitudes of the branch vectors [GUO 13].

The stress in a granular medium is the volume average of the dyadic products of branch vectors and contact forces, $\mathbf{l}^c \otimes \mathbf{f}^c = l_i^c f_j^c$, among the contacts within an assembly ($\mathbf{f}^c$ is the contact force, see section 1.3). Owing to the strong correlation between orientations of the contact forces and contact normals (since the differences in their directions is limited by the friction coefficient), the stress is strongly correlated with the average of these dyadic products, suggesting a third, mixed measure of contact orientation:

$$\overline{H}_{ij} = \frac{1}{M} \sum_{c=1}^{M} (l_i^c n_j^c)/|\mathbf{l}^c| \tag{1.19}$$

Evolution of the fabric tensors $\overline{\mathbf{F}}$, $\overline{\mathbf{G}}$ and $\overline{\mathbf{H}}$ during deviatoric loading is discussed in section 4.4.2.

Radjai *et al.* [RAD 98] observed that the contribution of a contact to a material's anisotropy is largely determined by the normal force that it carries. He proposed that contacts be divided into two categories: those that have a contact force greater than the mean force and those with a smaller force (section 4.5). Designating these groups as "strong" and "weak" contacts, they were found to have different distribution densities, and the strong contacts contributed disproportionately in bearing the deviatoric stress. These observations suggest a variation of the Satake tensor, by averaging the contact orientations among the subset of "strong" contacts:

$$\overline{F}_{ij}^{\text{strong}} = \frac{1}{M^{\text{strong}}} \sum_{c=1}^{M^{\text{strong}}} n_i^c n_j^c \tag{1.20}$$

in which the sum is confined to those contacts having a normal force larger than the average force. A weak counterpart $\overline{\mathbf{F}}^{\text{weak}}$ is similarly defined, and similar "strong" and "weak" forms of fabric tensors $\overline{\mathbf{G}}$ and $\overline{\mathbf{H}}$ can also be defined.

1.2.6. *Voids fabric*

Strain within a granular material is largely due to deformation of the void space: when the nearly rigid particles are rearranged, the pore space (or pore fluid) is altered in size and shape. As such, anisotropy of the pore space is indicative of an anisotropic grain arrangement that produces an anisotropic stiffness and a deviatoric stress. Moreover, anisotropy of the void space affects the hydraulic transport properties of granular, porous materials. Besides the importance of voids to the deformation and transport properties of a granular material, constrictions in the void space (i.e. pore throats) can block the migration of small particles that would otherwise be swept through the material by a flowing pore fluid. The size and geometry of the void space is critical in the design of soil filters and in the prevention of internal erosion (suffusion) in a soil. The void space can be characterized by the connectivity, size, shape and orientation of the voids, and suitable methods for describing these aspects in two- and three-dimensional systems are described in separate sections.

1.2.6.1. *Voids fabric in two dimensions*

The arrangement of particles and voids in two-dimensional assemblies is naturally captured in the particle (or Satake) graph, as shown in Figure 1.3(a), in which particles, contacts and voids are represented as nodes, edges and faces [SAT 93]. The graph not only provides a geometric partition (tessellation) of space but is also a succinct topologic portrayal of the material. In computations, the graph can be stored and rapidly traversed using a doubly connected edge list (DCEL, see [PRE 85]). For assemblies of non-convex particles, multiple contacts can exist between a pair of particles, and this topologic arrangement can be represented either with a simple graph in which multiple contacts are represented by a single edge or with a multigraph in which two nodes (particles) can share multiple edges. When represented with a simple graph, the numbers of nodes (particles) N, edges (contacts) M and faces (voids, loops or void cells) L are related by Euler's formula for a planar graph:

$$N - M + L = 2 \qquad\qquad [1.21]$$

In this version of the formula, the entire region outside of the assembly is considered a face, as if the graph were wrapped onto a sphere. If the exterior is not considered as a face, the 2 is replaced with 1. If the boundaries of a rectangular region are periodic, the 2 is replaced with zero (section 3.1.2). Unconnected, rattler particles are excluded from the graph (e.g. the shaded particle in Figure 1.3a). An assembly with non-convex particles and multiple contacts between particle pairs can be represented with a multigraph, with possibly multiple edges between particle nodes. To be consistent with equation [1.21], a double-contact between two particles is represented with two edges that form a two-edge loop, a triple-contact between two particles is represented by three edges that form two two-edge loops, etc.

As the coordination number of a single particle is its number of contacts, the *valence* of a single void is the number of particles that enclose it. The average valence $\overline{m}$ and average coordination number $\overline{n}$ are related by equations [1.15] and [1.21] as:

$$\overline{m} \approx 2\frac{M}{L} = 2 + \frac{4}{\overline{n} - 2} \qquad\qquad [1.22]$$

where the 2 in equation [1.21] has been ignored. Large average coordination numbers correspond to small average valences. The dual of a particle graph is the complementary void graph, in which the nodes and faces of a particle graph reverse their roles (Figure 1.3(b)), and the valence of a single void is the number of edges that are joined at the void's vertex.

An alternative to the particle graph is a tessellation proposed by Konishi and Naruse [KON 88], referred to as a K–N graph (Figure 1.3(c)). In this approach, the particles are represented by "solid" polygons, and voids are represented by loops that join the chords between contacts within the surrounding particles. Note that the edges of a particle graph become vertices in the K–N graph and that both vertices and faces in the particle graph become faces in the K–N graph. The K–N tessellation has the advantage of representing the region as a system of load-bearing sub-regions (the particle polygons) and deformation-producing sub-regions (the void polygons). This aspect allows the same graph to be used in computing stress and strain (e.g. [LI 09]).

Scalar measures of void fabric include the average valence $\overline{m}$ and the average void ratio e (section 1.1). The void ratio is widely used in geotechnical engineering as a measure of the compactness or "density" of soils. Spatial and probability distributions of local density and valence are measures of heterogeneity in granular materials [SHA 83, ANN 93]. The local density of physical samples is typically computed using image analysis and stereologic techniques [BHA 90, KUO 98, ALS 99], whereas the local void density in numerical simulations can be readily computed from the particles' geometric descriptions.

Three tensors have been proposed for characterizing the size and orientations of voids in two-dimensional systems. Noting that the edges of the particle graph are the branch vectors between particle centers (Figure 1.3(a)), Tsuchikura and Satake [TSU 01] computed the symmetric loop tensor $\mathbf{T}^\ell$ of a single void polygon ℓ as the sum of dyads of the individual branch vectors $\mathbf{l}$ that surround the loop (Figure 1.4),

$$T_{ij}^\ell = \frac{1}{2} \sum_\ell l_i l_j = \alpha^\ell \delta_{ij} + T_{ij}'^{,\ell} \qquad [1.23]$$

which is separated into isotropic and deviatoric parts: $\alpha^\ell = \frac{1}{2} T_{ii}^\ell$ and $T_{ij}'^{,\ell} = T_{ij}^\ell - \alpha^\ell \delta_{ij}$. Averaging the loop tensors of all void polygons in an assembly gives a measure of the size and orientation of the assembly's voids:

$$\overline{T}_{ij} = \overline{\alpha} \delta_{ij} + \overline{T}_{ij}' \qquad [1.24]$$

where $\overline{\alpha}$ is a rough measure of the average size of the void cells. The eigenvectors of $\overline{\mathbf{T}}'$ indicate the average elongated (and shortened) directions of the voids, and the ratio of its eigenvalues indicates the degree of elongation.

Kruyt and Rothenburg [KRU 96] introduced a polygon vector $\mathbf{h}^c$ associated with each edge (contact) c in a two-dimensional assembly (see Figure 1.9). Vector $\mathbf{h}^c$ is found by rotating the void vector $\mathbf{g}^c$, which connects the centers of the two void polygons that share the edge c, counterclockwise by 90°. Similar to the fabric tensor $\overline{\mathbf{G}}$ that was derived from the branch vectors joining particle centers in equation [1.18], a fabric tensor is computed from the polygon vectors of an assembly:

$$\overline{K}_{ij} = \frac{1}{M} \sum_{c=1}^{M} (h_i^c h_j^c)/|\mathbf{h}^c|^2 \qquad [1.25]$$

which is a measure of voids orientation. A probability density, similar to that in equation [1.17], can also be computed for tensor $\overline{\mathbf{K}}$.

Konishi and Naruse [KON 88] proposed a third fabric tensor of voids orientation, similar to the Tsuchikura and Satake tensor of equation [1.23], that is based on the chord vectors $\mathbf{v}^k$ that surround a single void ℓ (i.e. vector edges of the void polygons in Figure 1.3(c)):

$$P_{ij}^{\ell} = \frac{1}{2} \sum_{k} v_i^k v_j^k \qquad [1.26]$$

This tensor can be expressed in isotropic and deviatoric parts and averaged among all voids in an assembly (as with equations [1.23] and [1.24]).

1.2.6.2. Voids fabric in three dimensions

Many characteristics of the void space described for two-dimensional systems also apply in three dimensions: topology and connectivity, size, shape and orientation. The added dimension, however, introduces practical difficulties in characterizing and quantifying the void fabric. For example, the presence and location of contacts in a two-dimensional physical model can be easily distinguished in photographic or digital images, whereas the identification of contacts among three-dimensional particles can be difficult to surmise and may require advanced imaging methods. As such, we must distinguish between fabric characteristics that apply in principle and those that can be ascertained in practice. Moreover, the graph of nodes (particles) and edges (contacts) is not planar for three-dimensional assemblies, and other topologic objects must be considered so that two-dimensional concepts, such as Euler's formula, can be extended to three dimensions. This section describes methods for expressing three characteristics of the voids in three dimensions: connectivity, size and orientation.

In principle, the topologic connectivity of the void space $\mathcal{B}^v$ within an assembly $\mathcal{B}$ can be quantified with its Euler-Poincaré characteristic $\chi(\mathcal{B}^v)$

[MIC 01]. The Euler-Poincaré characteristic is a fundamental property of a solid and is one of the four geometric/topologic properties φ of a solid region P that possesses the "additive" property: $\varphi(P_1 \cup P_2) = \varphi(P_1) + \varphi(P_2) - \varphi(P_1 \cap P_2)$. The fact that two of these properties, volume and surface area, are of obvious importance suggests that the other two properties (the characteristic χ and the integral mean curvature) are also important in granular mechanics. The Euler-Poincaré characteristic of the void region is

$$\chi(\mathcal{B}^v) = \text{(no. connected regions, } K) + \text{(no. cavities, } C) - \text{(no. tunnels, } G) \quad [1.27]$$

where "no." means "number of". Since the void space between solid particles is entirely inter-connected, the number of connected regions K equals one. The cavities C ("holes") within the inter-connected void space are the isolated particles (rattlers) and the isolated particle clusters that are disconnected from other particles and fully surrounded by void space. As gravity will settle each particle against other particles, C is one for granular soils (i.e. a single connected particle network within the void space $\mathcal{B}^v$). In computer simulations that omit gravity, however, unconnected "rattler" particles can be present as well as numerous, and each rattler contributes to the number of cavities C. The number of tunnels in a 3D region (i.e. the genus $G(\mathcal{B}^v)$ of the void region) is a topological quantity: the maximum number of full cuts that can be made through $\mathcal{B}^v$ without producing any separated (void) regions.

With simple geometric objects, the quantities in equation [1.27] are fairly straightforward. For example, a hollow ball (sphere surface) has characteristic $\chi = 2$ ($K = C = 1$, $G = 0$); a solid sphere (ball) has characteristic 1 ($K = 1$, $C = G = 0$); a solid sphere pierced with one hole (i.e. a solid torus) has characteristic 0 ($K = G = 1$, $C = 0$); a solid sphere pierced with two non-intersecting holes has characteristic -1 and a solid sphere connected by two holes that pass fully through the body and intersect each other has characteristic -2 ($K = 1$, $C = 0$, $G = 3$). Quantifying χ and G for the void space of a granular material is more challenging but can be achieved with any of the following three methods, depending on the type of data that are available.

When digitized data are available, a three-dimensional region is represented with voxels arranged in a regular grid, and χ can be computed with methods of integral geometry and morphological image analysis [MIC 01]. The characteristic χ of, say, the black voxels representing void space $\mathscr{B}^v$ can be computed as the number K of regions of connected black voxels, plus the number of completely enclosed regions C of white (solids) voxels, and minus the number of regions G of white voxels piercing regions of connected black voxels. More precisely, by imagining each voxel as a cube, the characteristic is given by:

$$\chi = -n_c + n_f - n_e + n_v \qquad [1.28]$$

where n_c is the number of black cubes, n_f is the number of faces among the black cubes, n_e is the number of edges among the black cubes and n_v is the number of vertices among the black cubes [MIC 01]. Table 1.1 applies this approach to a Rubik's cube of 27 cubes (voxels). One shortcoming of digitized data is the difficulty of distinguishing individual contacts between particles. If the resolution is too coarse, solid contacts that should appear as white voxels will instead appear as black void spaces. The presence of such obscured contacts can be approximated by applying discrete morphological methods [SER 82], such as dilation, erosion, opening and closing. For example, repeated closing operations applied to seemingly isolated solid voxels can meld them together, so that the contact "bridges" between grains are distinguished. This method, however, leads to a loss in the resolution of the particle shapes.

	n_c	n_f	n_e	n_v	χ
Solid cube	27	108	144	64	1
Solid cube with enclosed central hole	26	108	144	64	2
Cube with one piercing tunnel	24	104	144	64	−1
Cube with two piercing tunnels	22	100	144	64	−2

Table 1.1. *The Euler-Poincaré characteristic χ of a Rubik's cube of 27 voxels, contained within a $4 \times 4 \times 4$ grid of vertices*

When the available data include the locations and geometric shapes of the particles, as with computer simulations, the characteristic χ and genus G can be derived by constructing the void connectivity graph in which pore bodies (represented as graph nodes) are connected through restricted passageways (pore throats, represented as graph edges) between the particles [DEH 72, KWI 90, REE 96, HIL 03]. This full graph of the void space can be represented as the reduced medial axis (skeleton or deformation retract) of the void space and has been used to model the flow of fluids through the inter-connected   voids   [SER 82,   LIN 96,   LIA 00].   Genus G (in equation [1.27]) of the void space is [ADL 92]:

$$G(\mathcal{B}^v) = 1 + (\text{no. pore throats}) - (\text{no. pore bodies}) \qquad [1.29]$$

When $K = 1$ and $C = 1$, a large positive genus G in equation [1.29] (or a large negative Euler-Poincaré characteristic $\chi(\mathcal{B}^v)$ in equation [1.27]) indicates that many redundant pathways (i.e. pore throats or tunnels) are available to promote fluid migration through the void space. Prasad *et al.* [PRA 91] present a corresponding formula for the genus of the solid phase.

Discrete element (DEM) and other numerical simulations allow for the direct computation of χ by applying Minkowski functionals to the geometric descriptions of particle shapes. Minkowski functionals (i.e. Minkowski scalars) arise in integral geometry as four independent scalar values, which include volume and surface area. The four scalars that are associated with a 3-dimensional (3D) geometric body possess the additive property described above equation [1.27] and are invariant with respect to translation or rotation of the body. The complete set of Minkowski v-functionals W_v of a three-dimensional object $\mathcal{B}$ are given in the top part of Table 1.2, adapted from the summary of Schröder-Turk *et al.* [SCH 11]. The table applies to a region $\mathcal{B}$ that is a finite union of convex (but possibly disconnected) objects. In the expressions, κ_1 and κ_2 are the principal curvatures of the object's surface ∂X, quantity $(\kappa_1 + \kappa_2)/2$ is the mean curvature and $\kappa_1 \kappa_2$ is the Gaussian curvature. Functional W_0 is the volume, W_1 is one-third of the surface area, W_2 is equal to $2\pi/3$ times the "mean breadth" $B(\mathcal{B})$ of the object, and W_3 is directly related to the characteristic χ (see equation [1.27]) as:

$$W_3 = \frac{4\pi}{3}\chi(\mathcal{B}) \qquad [1.30]$$

which is an expression of the Gauss-Bonnet formula. Evaluating the functionals W_2 and W_3 for shapes with sharp edges or corners can be accomplished with cylindrical or spherical rounding (thus creating a smooth, differentiable surface) and then finding the limits of the integrations as the radius is reduced to zero.

Functional W_3 is $4\pi/3$ for a solid ball, and W_3 is $8\pi/3$ for two disjoint balls ($\chi = 1$ and 2, respectively). If two balls are brought into contact, forming a finite contact area, W_3 is reduced from $8\pi/3$ to $4\pi/3$: the two spherical surfaces have a positive Gaussian curvature $\kappa_1\kappa_2$ and together contribute $8\pi/3$ to the integral, but the bridge between the two spheres has a negative curvature and contributes $-4\pi/3$. By extension, the Euler-Poincaré characteristic χ of an assembly of solid particles $X(\mathcal{B}^s)$ connected at their contacts is:

$$\chi(\mathcal{B}^s) = \frac{3}{4\pi} W_3 = 1 + (\text{no. particles}) - (\text{no. contacts between particles}) \quad [1.31]$$

Type	Symbol	Definition
Functionals	$W_0(\mathcal{B})$	$\int_{\mathcal{B}} dV$
	$W_1(\mathcal{B})$	$\frac{1}{3} \int_{\partial\mathcal{B}} dA$
	$W_2(\mathcal{B})$	$\frac{1}{3} \int_{\partial\mathcal{B}} \frac{1}{2}(\kappa_1 + \kappa_2)\, dA$
	$W_3(\mathcal{B})$	$\frac{1}{3} \int_{\partial\mathcal{B}} \kappa_1\kappa_2\, dA$
Tensors	$W_{1,ij}^{2,0}(\mathcal{B})$	$\frac{1}{3} \int_{\partial\mathcal{B}} x_i x_j\, dA$
	$W_{1,ij}^{0,2}(\mathcal{B})$	$\frac{1}{3} \int_{\partial\mathcal{B}} n_i n_j\, dA$
	$W_{3,ij}^{2,0}(\mathcal{B})$	$\frac{1}{3} \int_{\partial\mathcal{B}} \kappa_1\kappa_2\, x_i x_j\, dA$

Table 1.2. *Selected Minkowski functionals and tensors for 3D objects X, adapted from Schröder-Turk et al. [SCH 11]*

As the void space and solid space share a common surface, $\partial\mathcal{B}^s = \partial\mathcal{B}^v$, and have the same Gaussian curvature, the Euler-Poincaré characteristic of the void space is also:

$$\chi(\mathcal{B}^v) = \chi(\mathcal{B}^s) \quad [1.32]$$

This approach to quantifying χ for a numerical assembly of particles (or void space) involves simply counting the numbers of particles and contacts (as in equation [1.31]) and does not require the direct evaluation of surface integrals. The void connectivity is normalized as:

$$\overline{\chi}^v = \chi(\mathcal{B}^v)/N \quad [1.33]$$

by dividing by the number of particles N. Note that $\overline{\chi}^v$ in equation [1.31] is directly related to the average coordination number $\overline{n}$ of equation [1.15] and to the degree of structural redundancy of the particle network (see section 4.3) [THO 98, KRU 09].

Minkowski *tensors* provide measures of the shape and orientation of the void space. Schröder-Turk *et al.* [SCH 13, SCH 11] identify six rank-two Minkowski tensors that form a complete set of isometry covariant, additive and continuous functions for three-dimensional poly-convex bodies. Three of the six tensors are given in Table 1.2. The first tensor $\mathbf{W}_1^{2,0}$ is similar to tensor $\overline{\mathbf{J}}$ of equation [1.12], which expresses the average orientation of the particles; the second tensor $\mathbf{W}_1^{0,2}$ is similar to $\overline{\mathbf{S}}$ of equation [1.14], which expresses the orientation of particle surfaces. The final tensor is covariant with respect to translation and rotation, depends solely on the shape, size and connectivity of a 3D object, and can be directly evaluated from the geometric data of numerical simulations for either the solid or void regions, as the surface integral:

$$W_{3,ij}^{2,0}(\mathcal{B}) = \frac{1}{3} \int_{\partial \mathcal{B}} \kappa_1 \kappa_2 \, x_i x_j \, dA \tag{1.34}$$

The meaning of this tensor (and the corresponding functional W_3) is illustrated in Figure 1.5 for the cases of a rectangular block of size $2a_1 \times 2a_2 \times 2a_3$ and of a block pierced by a square tunnel of size $2a_3 \times 2c \times 2c$. The sides and edges of the block have zero Gaussian curvature and do not contribute to W_3 or to $W_{3,ij}^{2,0}$, whose values are derived entirely from the corners. The contribution to W_3 of a single corner is $\gamma/3$, where γ is its Descartes angular deficit ($\pi/2$ for a square corner). The value of W_3 for the eight corners of a solid block is $8(\pi/2)/3 = 4\pi/3$, corresponding to $\chi = 1$. If the coordinate system is centered within the block, tensor $\mathbf{W}_3^{2,0}$ is formed from the dyads $\mathbf{x} \otimes \mathbf{x} = x_i x_j$, where vectors x_i are directed from the block's center to its corners:

$$\mathbf{W}_3^{2,0}(X \text{ of Figure 1.5(a)}) = \frac{1}{3} 8 \frac{\pi}{2} \begin{bmatrix} a_1^2 & 0 & 0 \\ 0 & a_2^2 & 0 \\ 0 & 0 & a_3^2 \end{bmatrix} \tag{1.35}$$

This tensor captures information of the block's shape (Figure 1.5(a)). The values of $\mathbf{W}_3^{2,0}$ for the pierced block (Figure 1.5(b)) equals the expression in equation [1.35] plus contributions from the eight interior corners. Each of these corners has a negative angular deficit, $-\pi/2$, such that:

$$\mathbf{W}_3^{2,0}(X \text{ of Figure 1.5(b)}) = \frac{4\pi}{3} \begin{bmatrix} a_1^2 - c^2 & 0 & 0 \\ 0 & a_2^2 - c^2 & 0 \\ 0 & 0 & 0 \end{bmatrix} \tag{1.36}$$

A tunnel is seen to modestly reduce the first two diagonal terms, while reducing the third term to zero. If the solid region represents a granular material's grains, then multiple tunnels in the x_3 direction make the third term negative, an indicator of multiple pathways (and void anisotropy) in this direction.

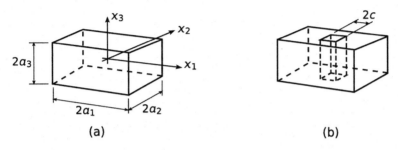

(a)   (b)

Figure 1.5. *Rectangular blocks as an illustration of Minkowski tensors: (a) solid block and (b) solid block with hole*

The tensor $\mathbf{W}_3^{2,0}$ of a single object $\mathcal{B}$ whose (local) center is offset by vector $\mathbf{t}$ from the origin O of a (global) coordinate frame is:

$$W_{3,ij}^{2,0} = \check{W}_{3,ij}^{2,0} + 2\,t_i\left(\frac{1}{3}\int_{\partial\mathcal{B}} \kappa_1\kappa_2\check{x}_j\,dA\right) + W_3\,t_i t_j \qquad [1.37]$$

(see equation 6 in [SCH 11]), which is the parallel-axis relationship for tensor $\mathbf{W}_3^{2,0}$. In this equation, $\check{\mathbf{W}}_3^{2,0}$ and $\check{x}_i$ are measured relative to the local center of the object, and the expression in parentheses is the local Minkowski vector $\check{W}_{3,i}^{1,0}$, which is equal to zero for spheres, ellipsoids and other objects having orthorhombic symmetry. Scalar W_3 is defined in Table 1.2.

For a granular assembly, the region in equations [1.34] and [1.37] can be taken as the solid particles $\mathcal{B}^s$, which are joined at their contacts. As with equation [1.31], the integration [1.34] around each particle makes a positive contribution to $\mathbf{W}_3^{2,0}$, whereas negative curvatures at the bridges (contacts) between particles make negative contributions to $\mathbf{W}_3^{2,0}$. Combining equations [1.34] and [1.37], noting that $W_3 = 4\pi/3$ for a solid particle without holes and that $W_3 = -4\pi/3$ for a contact bridge, and assuming that the vector $\check{W}_{3,i}^{1,0} = 0$ for each particle, we have:

$$W_{3,ij}^{2,0}(\mathcal{B}^s) = \sum_{p=1}^{N}\left(\check{W}_{3,ij}^{2,0,p} + \frac{4\pi}{3}x_i^p x_j^p\right) - \sum_{c=1}^{M}\frac{4\pi}{3}x_i^c x_j^c \qquad [1.38]$$

In this expression, contributions are summed for the N particles and the M contacts: $\check{W}_3^{2,0,p}$ is the local tensor for particle p, $\mathbf{x}^p$ is the vector from the assembly's center to a particle's center and $\mathbf{x}^c$ is the vector from the assembly's center to a contact. In the example given in Figure 1.5, we note that the magnitudes of the components of tensor $\mathbf{W}_3^{2,0}$ depend on the overall shape and size of a region $\mathcal{B}$ as well as on the connectivity within the region.

The tensor in equation [1.38] is quadratic in the components x_i^p and x_i^c of the particle locations and the contact locations. As such, the tensor $\mathbf{W}_3^{2,0}(\mathcal{B}^s)$ is sensitive to the overall size of an assembly, rather than to the local void fabric. The overall shape and size of the assembly will also change during deformation, masking the underlying changes in the fabric. To compensate, we divide $\mathbf{W}_3^{2,0}$ by the integral in equation [1.34], as applied to the full assembly's outer boundary:

$$\overline{W}_{3,ij}^{2,0}(\mathcal{B}^s) = (W_{3,ik}^{2,0})^{-1}_{\text{equation [1.34], boundary}}(W_{3,kj}^{2,0})_{\text{equation [1.38]}} \qquad [1.39]$$

thus giving a normalized measure of the orientations of tunnels (voids) within a granular region.

Equation [1.28] can be applied to digitized images of granular media to characterize its connectivity. Images can also be analyzed to find measures of the sizes, shapes and orientations of the voids. By applying mathematical morphology methods to three-dimensional digitized images, Hilpert [HIL 03] developed a method for estimating the cumulative distribution f_ρ^v of pore size ρ (see also [VOG 97]):

$$f_\rho^v(\rho) = \frac{\text{Vol}\left(O_\rho(X^v)\right)}{\text{Vol}(X^v)} \qquad [1.40]$$

which gives the fraction of the void volume (i.e. fraction of the void voxels) that fully encompasses a sphere of radius ρ. The void volume in the denominator is a simple counting of the number of void voxels in the image. The quantity $O_\rho(X^v)$ in the numerator is a counting of the morphological opening of X^v with a sphere-shaped structural template of 1's having radius ρ [SER 82]. A morphological opening (a dilation with the sphere-shaped template of radius ρ followed by an erosion with the same template) removes thin passageways and throats that are smaller than the template's size. As a limiting example, if the template is a single null voxel (representing radius $\rho = 0$), the opening operation (i.e. an erosion followed by a dilation) leaves the image unchanged, and the quotient is 1.0: 100% of the void space is larger than size 0. When the template is a digitized ball of radius ρ, the quotient is the fraction of void voxels that lie at a distance greater than ρ from the nearest particle. In this manner, the $f_\rho^v(\rho)$ represents the fraction of voids volume larger than the digitized radius ρ.

The use of a digitized sphere can be generalized to templates of other shapes. To capture the elongation and direction of the void space, we can use a "spar" of length ℓ_i oriented in the grid direction x_i: simply a single row of digital 1's of length ℓ along this direction. The cumulative distribution of path lengths in this direction can then be produced, giving a measure of the unobstructed lengths of pores in that direction [KUH 15]. Likewise, a digitized disk-shaped template perpendicular to direction x_i can be used to construct the cumulative distribution of passageways larger than a given size for flow in that direction.

Kuo *et al.* [KUO 98] used stereologic methods with photographic images of polished thin sections to compute a mean free path tensor $\overline{\lambda}$, characterizing the orientation density of voids. A continuous line with direction **n** will pass through both voids and grains, and the mean free path is the average of the uninterrupted lengths of line segments within the voids. By analyzing multiple directions and using the methods of section 1.2.3, a tensor $\overline{\lambda}$ can be computed to fit length $\lambda(\mathbf{n})$ to the direction of the mean path:

$$\lambda(\mathbf{n}) \approx \lambda_0 (1 + \overline{\lambda}_{ij} n_i n_j) \qquad [1.41]$$

where λ_0 is the average uninterrupted length of the pore space (see equation [1.2]). The median values of the three lengths l_i computed with equation [1.40], but with spars of lengths ℓ_i, represent the median free paths in directions x_1, x_2 and x_3. A matrix $\overline{\mathbf{L}}^v$ can be constructed from the distribution of these three lengths, with the diagonal elements:

$$\overline{L}_{ii}^v = \frac{3}{\overline{\ell}}\overline{\ell}_i \qquad\qquad [1.42]$$

where $\overline{\ell}_i$ is the median value of ℓ_i for which $f_{\ell_i}^v(\ell_i) = 0.50$, and $\overline{\ell}$ is the trace $\overline{\ell}_1 + \overline{\ell}_2 + \overline{\ell}_3$. This matrix is a measure of anisotropy in the orientation of the void space.

1.3. Granular stress

Stress usually connotes a force per unit area, taken at the limit of an infinitesimal area. Stress is a tensor, as the force per area depends on the area's orientation. We immediately acknowledge a question, however, when expressing stress within a discontinuous and inhomogeneous material: should an infinitesimal area be considered a mathematical point having zero measure or should the area (and the stress) have a finite but small scale? The discontinuous nature of granular materials gives rise to different definitions of stress, each useful in a particular setting. Moreover, stress can be generalized to include other force-like objects, leading to couple-stress and higher-order stress. In this section, we develop two definitions of stress: the average stress and several definitions of the continuum equivalent stress. Each definition will be seen to depend on the continuum setting in which it is applied: a classical continuum, a micro-polar continuum or other generalized continua. In the following, we depart from geotechnical convention, by adopting compressive stress as negative and tensile stress as positive.

1.3.1. *Average stress*

Recognizing that granular materials are composed of discrete solid grains and a fluid phase (or phases) within the pore space, a natural approach is to express the average stress within a single grain, the average stress within a region of grains and the average stress (pressure) within the pore space. If Ω^p is the material region of a single particle p, within which the Cartesian components of the Cauchy stress tensor σ can vary with position $\mathbf{x}$ within p, the particle's volume-average stress is

$$\overline{\sigma}_{ij}^{p} = \frac{1}{V^{p}} \int_{\Omega^{p}} \sigma_{ij}(\mathbf{x})\, dV \qquad [1.43]$$

where the particle's volume (or area, for 2D particles) is $V^{p} = \int_{\Omega^{p}} dV$. By applying the Gauss-Green principle, equation [1.43] can be written as:

$$\overline{\sigma}_{ij}^{p} = \frac{1}{V^{p}} \int_{\partial\Omega^{p}} x_{i}\sigma_{kj}n_{k}\, dS - \frac{1}{V^{p}} \int_{\Omega^{p}} x_{i}\sigma_{kj,k}\, dV \qquad [1.44]$$

According to Cauchy's first law, the divergence of stress in the second integral, $\vec{\nabla} \cdot \sigma = \sigma_{kj,k}$, is related to the body force density $\mathbf{b}$, the mass density ρ and the acceleration $d\mathbf{v}/dt$ as follows: $\sigma_{kj,k} + b_{j} = \rho\, dv_{j}/dt$. In the following, we will usually neglect the self-weight of the particles, any electrostatic interactions between the particles and their surroundings, and any inertial forces and accelerations, making the second integral zero. The first integral in equation [1.44] is across the particle's surface $\partial\Omega^{p}$ (or perimeter in 2D). The term $\mathbf{n} \cdot \sigma = \sigma_{kj}n_{k}$ is the externally applied surface traction, which can be produced by interactions with the neighboring particles and with the pore fluid. Static fluids will alter the average mean stress $\frac{1}{3}\overline{\sigma}_{kk}^{p}$ within particle p and will produce a buoyancy that can partially counteract its self-weight $\mathbf{b}$. Flowing fluids produce drag (seepage) forces and, when these forces fully counteract the self-weight, can lead to liquefaction and fluidization. Just as with the self-weight, we will often neglect the fluid pressures on particle p.

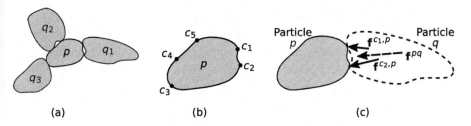

Figure 1.6. *Neighbors and forces of a particle p: (a) set of contacting particles $Q^{p} = \{q_{1}, q_{2}, q_{3}\}$; (b) set of contacts $C^{p} = \{c_{1}, c_{2}, c_{3}, c_{4}, c_{5}\}$ and (c) contact forces, $f^{c_{1},p}$ and $f^{c_{2},p}$, acting on p, and the resultant force f^{pq} of q on p.*

As depicted in Figure 1.6(a), a particle p can interact with the set of its neighboring particles, $Q^p = \{q_1, q_2, \ldots\}$. With non-convex particles, multiple contacts can exist between p and a neighbor q, such that p has the contact set $C^p = \{c_1, c_2, \ldots\}$ among its neighbors (Figure 1.6(b)). Inter-particle interactions will be limited to discrete, point-like contact forces through which a typical neighbor exerts force $\mathbf{f}^{c,p}$ on p at the single contact c. Contact moments $\mathbf{m}^{c,p}$ are neglected presently but are later considered in section 1.3.2. A neighboring particle q might touch p at multiple points, and $\mathbf{f}^{pq}$ represents the resultant on p of the forces at these multiple contacts (Figure 1.6(c)). In the absence of body forces $\mathbf{b}^p$, the resultant of the contact forces on p is zero,

$$\sum_{c \in C^p} f_i^{c,p} = \sum_{q \in Q^p} f_i^{pq} = 0 \qquad [1.45]$$

so that the choice of an origin of the global coordinate system is extraneous in the first integral of equation [1.44] (an expression of invariance and symmetry with respect to translations). The surface integral in equation [1.44] can be expressed as a sum of dyads $\mathbf{x}^c \otimes \mathbf{f}^{c,p}$,

$$\overline{\sigma}_{ij}^p = \frac{1}{V^p} \sum_{c \in C^p} x_i^c f_j^{c,p} \qquad [1.46]$$

where $\mathbf{x}^c$ is the location of the contact point c relative to the global origin O, as in Figures 1.4 and 1.7.

The volume-average stress within an assembly $\mathcal{B}$ can be found as the weighted sum of the average stresses among the assembly's particles (Figure 1.7(a)). The region $\Omega^{\mathcal{B}}$ is delineated by its boundary $\partial\Omega^{\mathcal{B}}$ and has volume $V^{\mathcal{B}}$. The assembly $\mathcal{B}$ could be a sub-region (meso-scale region) of a larger granular region or could be an entire self-contained specimen or region of particles. The average stress within the region is the volume average of the stress within the void and solid portions of $\Omega^{\mathcal{B}}$, the latter stress given for individual particles by equation [1.46]:

$$\overline{\sigma}_{ij}^{\mathcal{B}} = \frac{1}{V^{\mathcal{B}}} \left(V^{\mathcal{B},\text{voids}} \, \overline{\sigma}_{ij}^{\text{voids}} + V^{\mathcal{B},\text{grains}} \, \overline{\sigma}_{ij}^{\text{grains}} \right) \qquad [1.47]$$

$$= -\hat{p}\, \delta_{ij} + \frac{1}{V^{\mathcal{B}}} \sum_{p \in \mathcal{B}} \sum_{c \in C^p} x_i^c f_j^{c,p} \qquad [1.48]$$

where $\hat{p}$ is the average fluid pressure within the pore space of $\Omega^{\mathcal{B}}$ (i.e. the average negative mean stress, $-\frac{1}{3}\sigma_{kk}$, of the fluids). The Gauss–Green principle has been applied to the integral in equation [1.48], and the contributions of all particles within the region, $p \in \mathcal{B}$, have been summed. This average stress will depend, of course, on the volume $V^{\mathcal{B}}$ assigned to the assembly. The boundary $\partial\Omega^{\mathcal{B}}$ can either fully encompass the particles (as in Figure 1.7(a)), or it can be crafted to pass through the peripheral particles. The former is more appropriate when the assembly is a self-contained specimen, whereas the latter might be a more appropriate choice for assemblies that are sub-regions of larger assemblies. Expressions will be developed for the first type of boundary, as this choice avoids the need to evaluate stress at boundary points that lie inside the particles.

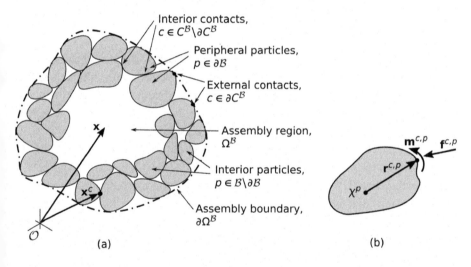

Figure 1.7. *Granular region $\mathcal{B}$: (a) interior and external contacts and (b) contact force and moment. Contact moments are considered in section 1.3.2*

Equation [1.48] shows that the average stress is the sum of the average pore pressure and the *inter-granular* stress (the effective stress) that is attributed to the contact forces among the particles, thus explicitly expressing the effective stress concept, a foundational principle in soil mechanics. Note that positive normal stress is tensile, whereas a positive pressure $\hat{p}$ is compressive. With this understanding, the equation equates the total stress with the sum of the pore pressure and the effective stress.

The summations in equation [1.48] can be replaced with either of two more succinct forms, by noting the self-equilibration of contact forces $\mathbf{f}^{pq} = -\mathbf{f}^{qp}$ and $\mathbf{f}^{c,p} = -\mathbf{f}^{c,q}$ (Newton's third law) and the equilibrium of each particle (as in equation [1.45]). The contributions of interior contact forces cancel, leaving only the external forces exerted on the peripheral particles:

$$\overline{\sigma}_{ij}^{\mathcal{B}} = -\hat{p}\,\delta_{ij} + \frac{1}{V^{\mathcal{B}}} \sum_{p\in\partial\mathcal{B}} \sum_{c\in\partial C^p} x_i^c f_j^{c,p} \qquad [1.49]$$

where the summations are confined to the peripheral particles p along the assembly's boundary, $p \in \partial\mathcal{B}$, and to the contacts between these boundary particles and the region's exterior, $c \in \partial C^p$. That is, the summations in equation [1.49] are limited to the external contacts of region $\Omega^{\mathcal{B}}$ (Figure 1.7(a)).

In a second form, equation [1.46] is applied to individual particles, but with a coordinate origin that is local to each particle: for example, the reference points χ^p and χ^q of particles p and q (Figure 1.4). These reference points can be placed at the centers of the particles or at other points either inside or outside of a particle. Position $\mathbf{x}^c$ relative to O is replaced with the contact vector $\mathbf{r}^{c,p}$ of contact c relative to the reference point χ^p of particle p (Figures 1.4 and 1.7(b)). As the contact forces are self-equilibrating, $r_i^{c,p} f_j^{c,p} + r_i^{c,q} f_j^{c,q} = (r_i^{c,p} - r_i^{c,q}) f_j^{c,p}$. The branch vector $\mathbf{l}^{pq}$ is defined as the location of the reference point χ^q of q relative to χ^p: that is, $\mathbf{l}^{pq} = \mathbf{r}^{c,p} - \mathbf{r}^{c,q} = \mathbf{x}^q - \mathbf{x}^p$. With this notation, we can replace the nested sums in equation [1.48] with a single summation, taken over the set of all contacts $C^{\mathcal{B}}$ within $\Omega^{\mathcal{B}}$ rather than only those particles on the boundary, as in equation [1.49]:

$$\sum_{p\in\mathcal{B}} \sum_{c\in C^p} x_i^c f_j^{c,p} = \sum_{p\in\mathcal{B}} \sum_{c\in C^p} r_i^{c,p} f_j^{c,p} = \sum_{c\in C^{\mathcal{B}}} l_i^c f_j^c \qquad [1.50]$$

where $\mathbf{x}^c$ has been replaced by $\mathbf{x}^p + \mathbf{r}^{c,p}$, and the equilibrium equation [1.45$_1$] has been applied to each inner sum of equation [1.50$_1$]. Branch vector $\mathbf{l}^c = \mathbf{l}^{pq}$ is directed from one of the contact's particles, say p, to the other particle, say q; and the corresponding force $\mathbf{f}^c = \mathbf{f}^{c,p}$ is exerted on p by q. Combining equations [1.49] and [1.50] yields the following expression for the average stress in $\Omega^{\mathcal{B}}$:

$$\overline{\sigma}_{ij}^{\mathcal{B}} = -\hat{p}\,\delta_{ij} + \frac{1}{V^{\mathcal{B}}} \sum_{p\in\mathcal{B}} \sum_{c\in C^p} r_i^{c,p} f_j^{c,p} = -\hat{p}\,\delta_{ij} + \frac{1}{V^{\mathcal{B}}} \sum_{c\in C^{\mathcal{B}}} l_i^c f_j^c \qquad [1.51]$$

where each contact c is included only once in the final summation, rather than being included twice, as in the intermediate double sum (since both p and q must be considered in its final sum, both contributing to a single contact c). Note that two spherical particles, when pressed together, have a compressive contact force $\mathbf{f}^c$ that is aligned with $\mathbf{l}^c$ but is in the opposite direction of $\mathbf{l}^c$, which produces a negative inner product $l_i^c f_i^c$ and a negative (compressive) contribution to stress.

Along the region's boundary contacts $\partial C^{\mathcal{B}}$, external forces act upon the peripheral particles of region $\Omega^{\mathcal{B}}$. Noting that the boundary passes through the contact points and not through the peripheral particles themselves (consistent with Figure 1.7(a)), the boundary contact forces must be included among the terms $x_i^c f_j^{c,p}$ and $r_i^{c,p} f_j^{c,p}$ of equations [1.50] and [1.51] and within the separate quantities $l_i^c f_j^c$ in equation [1.51]. The "branch" vector of the exterior contacts is the vector $\mathbf{r}^{c,p}$ directed from the particle's reference point χ^p to the boundary contact. The interior and peripheral quantities are as follows:

$$\mathbf{l}^c = \begin{cases} \mathbf{x}^q - \mathbf{x}^p & c \in C^{\mathcal{B}} \backslash \partial C^{\mathcal{B}} \\ \mathbf{x}^c - \mathbf{x}^p & c \in \partial C^{\mathcal{B}} \end{cases} \qquad \mathbf{f}^c = \begin{cases} \text{interior force on } p & c \in C^{\mathcal{B}} \backslash \partial C^{\mathcal{B}} \\ \text{exterior force on } p & c \in \partial C^{\mathcal{B}} \end{cases} \qquad [1.52]$$

where $\mathbf{x}^c$ is the position of the exterior contact point $c \in \partial C^{\mathcal{B}}$ (i.e. its position relative to origin O, Figure 1.7(a)). With this understanding, the four equations, [1.47], [1.48], [1.49] and [1.51], are equivalent expressions for the average stress within a granular region. We will usually neglect the pore pressure, so that $\overline{\sigma}^{\mathcal{B}}$ represents the inter-granular, effective stress, given as:

$$\overline{\sigma}_{ij}^{\mathcal{B}} = \frac{1}{V^{\mathcal{B}}} \sum_{p \in \mathcal{B}} \sum_{c \in C^p} x_i^c f_j^{c,p} = \frac{1}{V^{\mathcal{B}}} \sum_{p \in \partial \mathcal{B}} \sum_{c \in \partial C^p} x_i^c f_j^{c,p} = \frac{1}{V^{\mathcal{B}}} \sum_{c \in C^{\mathcal{B}}} l_i^c f_j^c \qquad [1.53]$$

with negative values for compressive normal stress components.

1.3.2. Continuum equivalent stress

Stress is a continuum concept, with the stress field expressed at idealized infinitesimal points within a spatial continuum. When modeling stress and deformation within a large region – within an embankment, under a foundation or around an excavation – we commonly treat the granular medium as a continuous material (perhaps with a finite element model), eschewing its discrete, discontinuous nature at the particle scale. Many models treat the underlying medium as a classical continuum in which all relevant kinematics are captured by the displacement or velocity field, and in which stress depends exclusively on the spatial gradients of displacement or velocity. As will be seen, the stress in a classical continuum has a strictly "force" character, the stress tensor is symmetric and couple-stress is disallowed. However, in Cosserat and other generalized continua, the displacement or velocity field is augmented with other kinematic fields, such as with a micro-rotation field of material points that is independent of the skew-symmetric part of the displacement gradient or with a micro-stretch field or other enhanced kinematic information. These augmented continua are intended to capture aspects of material behavior that arise from a micro-structure that is discontinuous or non-smooth, as with the discrete, particulate nature of granular materials. This discrete discontinuous aspect of granular materials is thought by some investigators to control the thickness of shear bands and other phenomena that are difficult to explain in the context of a classical continuum of simple material.

Regardless of the particular continuum representation of a granular medium, we often wish to extract values of stress and deformation – objects that are associated with idealized points – from a finite region that contains a multitude of discrete grains of finite size. A *representative volume* or representative volume element (RVE) of granular material is an assembly with a sufficiently large number of grains to capture the behavior of a material, so that continuum measures of stress, strain and stiffness can be extracted. Ideally, the size of the RVE is sufficiently small, so that its behavior can be efficiently determined through computational, analytic or experimental means. In a sense, the representative volume serves as a "test assembly" of particles for estimating the stress–strain behavior of a much larger region of material, so that the larger region can be treated as a continuum.

The principle of virtual work provides a natural means of transitioning between discrete and continuous measures of internal force and moment within a representative volume: for example, from discrete internal contact forces and moments to an internal stress and couple-stress [CHA 06]. Virtual work is often viewed as an alternative formulation (a so-called "weak form") of the equilibrium (Newton's) equations or the equations of momentum conservation, which are the "strong form" equations. Paul Germain [GER 73] expounded an alternative view of virtual work by noting that force and stress are more abstract notions than movement and strain, and the virtual work principle enables definitions of force and stress vis-à-vis their more tangible counterparts. In its most fundamental form, the principle of virtual displacements holds that the work attributed to the virtual motions of external forces within a region plus the work attributed to an admissible field of internal virtual movements or generalized strains must equal zero:

$$\delta W^{Ext}(\mathbf{f}^{Ext}, \delta\mathbf{u}^{Ext}) + \left\{ \begin{array}{c} \delta W^{Int}(\mathbf{f}^{Int}, \delta\mathbf{u}^{Int}) \\ \text{or} \\ \delta W^{Int}(\widehat{\sigma}^{Int}, \delta\varepsilon^{Int}) \end{array} \right\} = 0 \qquad [1.54]$$

as in [GER 73]. Both cases of this equation are legitimate expressions of the virtual displacement principle. The $\mathbf{f}^{Ext}$ are generalized external forces (tractions and couple-tractions, body forces and body moments, etc.), and the $\mathbf{f}^{Int}$ are the internal forces that operate between particles (contact forces, contact moments, etc.). The internal virtual movements , $\delta\mathbf{u}^{Ext}$ and $\delta\mathbf{u}^{Int}$, are work-conjugate with their corresponding forces, whereas the internal virtual macro-strains $\delta\varepsilon^{Int}$ are conjugate with certain internal *macro-stresses*, $\widehat{\sigma}^{Int}$ (stress, couple-stress, stress-moment, etc.). By assuming that the two cases of internal work in equation [1.54] are equivalent, we can derive expressions for the more opaque macro-stresses $\widehat{\sigma}^{Int}$ in terms of the more overt internal contact forces and moments, $\mathbf{f}^{Int}$ and $\mathbf{m}^{Int}$.

The virtual movements and strains are hypothetical, rather than actual, quantities, and the enforcement of equations [1.54] establishes equilibrium between the external and internal forces and macro-stresses. The internal virtual movements and strains, $\delta\mathbf{u}^{Int}$ and $\delta\varepsilon^{Int}$, are arbitrary, provided that they are kinematically consistent with each other and with the external virtual movements $\delta\mathbf{u}^{Ext}$ and with any external or internal constraints that apply to the region (external supports, internal constraints such as the isochoric undrained condition, rigid grains, etc.). When applied in the setting of a classical continuum, equation [1.54] leads to the standard, strong form of equilibrium (i.e. Cauchy's equations):

$$\sigma_{ji,j} + b_i = 0 \quad \text{in } \Omega^{\mathcal{B}} \qquad [1.55]$$

$$\sigma_{ji}n_j = t_i \quad \text{on } \partial\Omega^{\mathcal{B}} \qquad [1.56]$$

where **b** and **t** are the body force density and surface traction (force per unit area). When applied in a continuum Cosserat setting, the equation yields the strong form of micro-polar equilibrium:

$$\left.\begin{array}{l} \sigma_{ji,j} + b_i = 0 \\ \mu_{ji,j} + e_{ijk}\sigma_{jk} + c_i = 0 \end{array}\right\} \text{ in } \Omega^{\mathcal{B}} \qquad [1.57]$$

$$\left.\begin{array}{l} \sigma_{ji}n_j = t_i \\ \mu_{ji}n_j = z_i \end{array}\right\} \text{ on } \partial\Omega^{\mathcal{B}} \qquad [1.58]$$

where μ is the couple-stress and c and z are the body-couple density and moment traction [ERI 67, MAL 69]. By including the tensor μ with its additional units of length, a micro-polar continuum introduces a length-scale to boundary value problems.

Although equations [1.54] express the equilibrium of external forces, the top part, when applied to a discrete material, also gives the equilibrium relationship between the discrete external forces and the discrete internal forces (e.g. equivalence of the left and right parts of equations [1.49] and [1.51]). The bottom part provides rational expressions for the internal stress and couple-stress in terms of the discrete internal contact forces. We now apply equations [1.54] in the context of a micro-polar (Cosserat) continuum to determine the equilibrium relationship between the internal and external contact forces (and moments) and expressions for the stress and couple-stress in terms of these forces. As in Figure 1.7(b), contact moments **m**^{c} are now allowed at the contacts between particles. To simplify matters, however, we will neglect pore pressures, body forces and body-couples.

To expedite the analysis, we begin with the equations of equilibrium for a single particle p in terms of the contact forces and moments among its contact set C^p [CHA 06]:

$$\sum_{c \in C^p \setminus \partial C^{\mathcal{B}}} f_i^{c,p} + \sum_{c \in C^p \cap \partial C^{\mathcal{B}}} f_i^{c,p} = 0 \qquad [1.59]$$

$$\sum_{c \in C^p \setminus \partial C^{\mathcal{B}}} (m_i^{c,p} + e_{ijk}r_j^{c,p}f_k^{c,p}) + \sum_{c \in C^p \cap \partial C^{\mathcal{B}}} (m_i^{c,p} + e_{ijk}r_j^{c,p}f_k^{c,p}) = 0 \qquad [1.60]$$

The first equation is the same as the previous equation [1.45], except that we now distinguish between the external contacts of p, which lie on the boundary $\partial \Omega^{\mathcal{B}}$ of the assembly $\mathcal{B}$, and the interior contacts, which lie within the assembly's interior, $\Omega^{\mathcal{B}} \setminus \partial \Omega^{\mathcal{B}}$ (see Figure 1.7). This distinction will be essential in applying virtual displacements to boundary and interior points. The first summations in equations [1.59] and [1.60] are over all those contacts of p (set C^p) that are not among the set of external contacts of region $\Omega^{\mathcal{B}}$ (the set $C^{\mathcal{B}} \setminus \partial C^{\mathcal{B}}$). The second sums are over all exterior contacts of p: that is, the contacts of p (set C^p) that are also among the set of external contacts $\partial C^{\mathcal{B}}$.

We assume that the grains are rigid but with deformable contacts, and we consider a hypothetical, virtual displacement and rotation for each grain p: the virtual vectors $\delta \mathbf{u}^p$ and $\delta \omega^p$, with components δu_i^p and $\delta \omega_i^p$. A grain's rotation may take place about any reference point within the grain or even about a point outside of the grain: the assigned reference points χ^p (Figure 1.4). For a pair of contacting particles, their virtual movements will produce a virtual shifting of the two grains at their contact (i.e. a relative displacement of the surfaces of the two grains) and a relative rotation of the two grains at their contact. These virtual relative movements will be seen to produce virtual work when multiplied by the contact force and the contact moment.

As the grains are assumed to be rigid, internal deformation occurs only at the contacts (via the relative particle movements) and not within the rigid grains. The principle of virtual displacements relies on the chosen virtual displacements and rotations being as general as possible, provided, however, that the internal and external displacements are consistent. Accordingly, the virtual displacements of the exterior contacts can be *arbitrarily* chosen, even when this implies deformation of a peripheral grain between its exterior contacts and its reference point. If we were to restrict exterior contact points to move in a rigid manner about the particles' reference points, inconsistencies will result in the expressions of stress and couple-stress.

We now multiply the equilibrium equation [1.59] of the single particle p by the virtual displacement components δu_i^p and multiply equation [1.60] by the virtual rotation components $\delta \omega_i^p$. These displacements and rotations are of the particle reference points χ^p. The two products are added together, and the equations of all particles of assembly $\mathcal{B}$ are summed to produce the total virtual work, which, according to equation [1.54], must equal zero:

$$\delta W = \delta W^{\text{Ext}} + \delta W^{\text{Int}} = 0 \qquad\qquad [1.61]$$

$$\delta W = \sum_{p \in \mathcal{B}} \left(\sum_{c \in C^p \setminus \partial C^{\mathcal{B}}} f_i^{c,p} + \sum_{c \in C^p \cap \partial C^{\mathcal{B}}} f_i^{c,p} \right) \delta u_i^p \qquad\qquad [1.62]$$

$$+ \sum_{p \in \mathcal{B}} \left(\sum_{c \in C^p \setminus \partial C^{\mathcal{B}}} (m_i^{c,p} + e_{ijk} r_j^{c,p} f_k^{c,p}) + \sum_{c \in C^p \cap \partial C^{\mathcal{B}}} (m_i^{c,p} + e_{ijk} r_j^{c,p} f_k^{c,p}) \right) \delta \omega_i^p$$

$$+ \sum_{p \in \mathcal{B}} \left(\sum_{c \in C^p \cap \partial C^{\mathcal{B}}} (f_i^{c,p} - f_i^{c,p}) \right) \delta u_i^c + \sum_{p \in \mathcal{B}} \left(\sum_{c \in C^p \cap \partial C^{\mathcal{B}}} (m_i^{c,p} - m_i^{c,p}) \right) \delta \omega_i^c$$

$$= 0$$

where $\mathcal{B}$ is the set of assembly particles, $\mathcal{B} = \{p_1, p_2, \ldots\}$. The total virtual work is the sum of external and internal virtual works (equation [1.61]), which will be separated below. If equation [1.62] is assumed to be true for arbitrary δu_i^p and $\delta \omega_i^p$, the strong equilibrium equations [1.59] and [1.60] for individual grains are readily recovered. For example, letting $\delta u_1^p = 1$ and all other virtual displacements equal to zero, we recover the equation of force equilibrium in direction x_1 for particle p. The last two sums of equation [1.62] are identically zero and involve the displacements and rotations of exterior points, $\delta \mathbf{u}^c$ and $\delta \omega^c$. Although they are zero, the inclusion of these null sums will allow the work of external (boundary) forces to be separated from the internal work that is due to any deformation within the region $\Omega^{\mathcal{B}}$. The sums in equation [1.62] involve the virtual movements $\delta \mathbf{u}^p$ and $\delta \omega^p$ of the particles' reference points χ^p and the virtual movements $\delta \mathbf{u}^c$ and $\delta \omega^c$ of the particles' boundary points (note the "p" and "c" superscripts, which are described below). The separation is as follows:

$$\delta W^{\text{Ext}} = \sum_{p \in \mathcal{B}} \left(\sum_{c \in C^p \cap \partial C^{\mathcal{B}}} f_i^{c,p} \delta u_i^c + \sum_{c \in C^p \cap \partial C^{\mathcal{B}}} m_i^{c,p} \delta \omega_i^c \right) \qquad\qquad [1.63]$$

$$\delta W^{\text{Int}} = - \sum_{p \in \mathcal{B}} \left(\sum_{c \in C^p \setminus \partial C^{\mathcal{B}}} \left[f_i^{c,p} (\delta u_i^p - e_{ijk} r_j^{c,p} \delta \omega_k^p) + m_i^{c,p} \delta \omega_i^p \right] \right) \qquad\qquad [1.64]$$

$$+ \sum_{p \in \mathcal{B}} \left(\sum_{c \in C^p \cap \partial C^{\mathcal{B}}} \left[f_i^{c,p} \left(\delta u_i^p - (\delta u_i^c - e_{ijk} r_j^{c,p} \delta \omega_k^p) \right) + m_i^{c,p} (\delta \omega_i^c - \delta \omega_i^p) \right] \right)$$

recalling that contacts $c \in C^p \cap \partial C^B$ are exterior contacts, and contacts $c \in C^p \setminus \partial C^B$ are interior contacts. The "p" movements are of the particles' reference points; the "c" movements are of the exterior contact points. As such, the virtual external work in equation [1.63] results from movements of the exterior (boundary) points, whereas the two lines of internal virtual work in equation [1.64] result from two types of movements and deformations. The first line gives the work due to the relative movements of particles at their interior contacts. The second line in equation [1.64] is virtual work done in deforming the peripheral particles due to any mismatch between movements of the exterior contact points (movements $\delta \mathbf{u}^c$ and $\delta \omega^c$) and movements of the peripheral particles' reference points χ^p (movements $\delta \mathbf{u}^p$ and $\delta \omega^p$). That is, some internal deformation of grains may result from an arbitrary choice of the exterior and interior contact movements, and we must account for the resulting (virtual) internal work among the peripheral particles.

Stress is a continuum concept, and to make the link between a discrete force system and its continuum equivalent, we choose discrete virtual displacements that are consistent with a continuous virtual displacement field (see [CHA 06]). The particle movements and rotations are chosen to comply with the truncated Taylor series:

$$\delta u_i^p = \delta \overline{u}_i^1 + \delta \overline{u}_{ij}^2 x_i^p + \frac{1}{2} \delta \overline{u}_{ijk}^3 x_i^p x_j^p \qquad [1.65]$$

$$\delta \omega_i^p = \delta \overline{\omega}_i^1 + \delta \overline{\omega}_{ij}^2 x_i^p \qquad [1.66]$$

The overlines are used to denote virtual *macro-strains*, meaning strains that apply to the entire assembly, rather than to individual particles and their immediate vicinities. Taken together, these virtual macro-strains produce a continuous virtual displacement field. The macro-strains are work-conjugate with certain complementary *macro-stresses*, and these macro-stresses will be derived through the principle of virtual work. The numbers of subscripts and the overline numbers in the superscripts – one, two or three – distinguish the different orders of virtual movements $\delta \overline{u}$. The quantities $\overline{\omega}_i^1$ and $\overline{\omega}_{ij}^2$ are rotation fields that produce a regular pattern of rotations of the individual particles. As in a Cosserat continuum, the *micro-rotation* of a particle corresponding to this rotation field can differ from the skew-symmetric parts of the displacement gradient $\delta \overline{u}_{ij}^2$ and of the higher-order gradient $\delta \overline{u}_{ijk}^3$. Quantities $\delta \overline{u}_{ij}^2$, $\delta \overline{u}_{ijk}^3$ and $\delta \overline{\omega}_{ij}^2$ are, respectively, the virtual macro-strain (possibly non-symmetric), the virtual macro-strain gradient and the virtual macro-rotation gradient. Besides applying these fields to the particles' reference points χ^p, equations [1.65] and [1.66] are also applied to movements of the exterior contacts, $\delta \mathbf{u}^c$ and $\delta \omega^c$.

By applying the virtual displacements of equations [1.65] and [1.66], the external and internal virtual works of equations [1.63] and [1.64] are

$$\delta W^{\text{Ext}} = \sum_{c \in \partial C^{\mathcal{B}}} \left(\delta \overline{u}_i^1 f_i^c + \delta \overline{u}_{ij}^2 f_i^c x_j^c + \frac{1}{2} \delta \overline{u}_{ijk}^3 f_i^c x_j^c x_k^c + \delta \overline{\omega}_i^1 m_i^c + \delta \overline{\omega}_{ij}^2 m_i^c x_j^c \right) \tag{1.67}$$

$$\delta W^{\text{Int}} = \sum_{c \in C^{\mathcal{B}}} \left\{ (\delta \overline{u}_{ij}^2 + e_{ijk} \delta \overline{\omega}_k^1) f_i^c l_j^c + \frac{1}{2} (\delta \overline{u}_{ijk}^3 + e_{ij\ell} \delta \overline{\omega}_{\ell k}^2) f_i^c J_{jk}^c \right. \tag{1.68}$$

$$\left. + \delta \overline{\omega}_{ij}^2 \left[m_i^c l_j^c + e_{ik\ell} (\frac{1}{2} J_{\ell j}^c - x_\ell^c l_j^c) \right] \right\}$$

The summation of the first expression is over the set of exterior contacts $\partial C^{\mathcal{B}}$ along the boundary of $\Omega^{\mathcal{B}}$; whereas, the summation of equation [1.68] is over *all* contacts, both interior and exterior, in the assembly set $C^{\mathcal{B}}$. As in the derivation of equation [1.51] of the previous section, we have combined complementary terms $\mathbf{r}_i^{c,p}$ and $\mathbf{r}_i^{c,q}$ and the action–reaction relations $\mathbf{f}_i^{c,p} = -\mathbf{f}_i^{c,q}$ and $\mathbf{m}_i^{c,p} = -\mathbf{m}_i^{c,q}$ for the interior contacts. In the case of interior contacts, the expression δW^{Int} involves contact quantities (e.g. $\mathbf{f}^c$ and $\mathbf{J}^c$, defined below) that are referenced to one or the other (but not both) of the particles, say p or q. For example, if $\mathbf{f}^c$ is a force exerted on p by q, then $\mathbf{l}_i^c$ is the branch vector from p to q (Figures 1.4 and 1.6). For the exterior contacts in equation [1.68], the forces and moments are exerted on the peripheral particles, and position vectors are referenced to a peripheral particle p and directed toward the exterior contact at location $\mathbf{x}^c$. This distinction between interior and exterior contacts is summarized as follows:

$$l_i^c = \begin{cases} x_i^q - x_i^p & c \in C^{\mathcal{B}} \backslash \partial C^{\mathcal{B}} \\ x_i^c - x_i^p & c \in \partial C^{\mathcal{B}} \end{cases} \qquad J_{ij}^c = \begin{cases} x_i^q x_j^q - x_i^p x_j^p & c \in C^{\mathcal{B}} \backslash \partial C^{\mathcal{B}} \\ x_i^c x_j^c - x_i^p x_j^p & c \in \partial C^{\mathcal{B}} \end{cases} \tag{1.69}$$

$$f_i^c = \begin{cases} f_i^{c,p} = -f_i^{c,q} & c \in C^{\mathcal{B}} \backslash \partial C^{\mathcal{B}} \\ f_i^{c,p} & c \in \partial C^{\mathcal{B}} \end{cases} \qquad m_i^c = \begin{cases} m_i^{c,p} = -m_i^{c,q} & c \in C^{\mathcal{B}} \backslash \partial C^{\mathcal{B}} \\ m_i^{c,p} & c \in \partial C^{\mathcal{B}} \end{cases} \tag{1.70}$$

where the $\mathbf{x}^c$ of exterior contacts are the locations of the exterior contact points. These conventions are an extension of those in equation [1.52].

The virtual macro-strains in equations [1.65]–[1.68] have arbitrary values, and we derive equilibrium relationships between internal and external force quantities by setting a single strain component equal to 1 and the others to 0. This principle is expressed in the upper part of equation [1.54]. As the virtual macro-displacement $\delta \overline{u}_i$ does no internal work (it is absent in equation [1.68]), the corresponding sum in equation [1.68] is zero, and the sum of external forces $\mathbf{f}^c$ must be equal to zero, as expected. The other equivalences in equations [1.67] and [1.68], lead to the following equilibrium relations between internal and external forces:

$$\frac{1}{V^{\mathcal{B}}} \sum_{c \in \partial C^{\mathcal{B}}} f_i^c x_j^c = \frac{1}{V^{\mathcal{B}}} \sum_{c \in C^{\mathcal{B}}} f_i^c l_j^c \qquad [1.71]$$

$$\frac{1}{V^{\mathcal{B}}} \sum_{c \in \partial C^{\mathcal{B}}} f_i^c x_j^c x_k^c = \frac{1}{V^{\mathcal{B}}} \sum_{c \in C^{\mathcal{B}}} f_i^c J_{jk}^c \qquad [1.72]$$

$$\frac{1}{V^{\mathcal{B}}} \sum_{c \in \partial C^{\mathcal{B}}} m_i^c = -\frac{1}{V^{\mathcal{B}}} \sum_{c \in C^{\mathcal{B}}} e_{ijk} f_k^c l_j^c \qquad [1.73]$$

$$\frac{1}{V^{\mathcal{B}}} \sum_{c \in \partial C^{\mathcal{B}}} m_i^c x_j^c = \frac{1}{V^{\mathcal{B}}} \sum_{c \in C^{\mathcal{B}}} \left[m_i^c l_j^c - e_{ik\ell} f_k^c (J_{\ell j}^c - x_\ell^c l_j^c) \right] \qquad [1.74]$$

The factor $1/V^{\mathcal{B}}$ is included so that they can be used directly in the expressions for macro-stress derived below.

The macro-stress quantities that correspond to the interior, inter-particle forces depend on the chosen underlying continuum and its corresponding characterization of internal virtual work. With a classical continuum, the internal work density is

$$\delta W^{\text{Int,I}} = \widehat{\sigma}_{ji}^{\text{I}} \delta \overline{u}_{i,j}^2 \qquad [1.75]$$

which is designated as the first "I" expression of work density. Equating the volume integral of this form of internal work with the work of the externally applied forces constitutes a weak form of equilibrium. The corresponding strong form is given by equations [1.55]–[1.56], and the macro-stress $\widehat{\sigma}^{\text{I}}$ is simply the Cauchy stress σ [GER 73]. Note that the macro-strain field of equation [1.65] has been truncated at the $\delta\overline{u}_{ij}^2$ term, and the rotations $\delta\overline{\omega}_i^1$ are discounted entirely. The correspondence of the $\delta\overline{u}_{ij}^2$ terms in equations [1.67], [1.68] and [1.75] for arbitrary $\delta\overline{u}_{ij}^2$ yields expressions for the macro-stress $\widehat{\sigma}^I$, given as the two forms in equation [1.71], the latter being the same as the average stress derived in section 1.3.1 (see equation [1.53]).

A second form of internal virtual work involves a richer (but more restrictive) kinematic field of virtual displacements and rotations:

$$\delta W^{\text{Int,II}} = \widehat{\sigma}_{ji}^{II}(\delta\overline{u}_{ij}^2 - e_{ijk}\delta\overline{\omega}_k^1) + \widehat{\mu}_{ji}^{II}\delta\overline{\omega}_{ij}^2 \qquad [1.76]$$

in which we have added virtual rotations $\delta\overline{\omega}_k^1$ and rotation gradients $\delta\overline{\omega}_{ij}$. This virtual work is that of a Cosserat continuum, and the strong equations [1.57] and [1.58] are derived from the weak form of equation [1.76]. The macro-stresses are:

$$\widehat{\sigma}_{ji}^{II} = \frac{1}{V}\sum_{c\in\partial C^B} f_i^c x_j^c = \frac{1}{V}\sum_{c\in C^B} f_i^c l_j^c \qquad [1.77]$$

$$\widehat{\mu}_{ji}^{II} = \frac{1}{V}\sum_{c\in\partial C^B} m_i^c x_j^c = \frac{1}{V}\sum_{c\in C^B}\left[m_i^c l_j^c - e_{ik\ell}f_k^c(J_{\ell j}^c - x_\ell^c l_j^c)\right] \qquad [1.78]$$

The stress $\widehat{\sigma}^{II}$ is identical to that of a classical continuum, $\widehat{\sigma}^I$, but by augmenting the classical continuum field with the rotations $\overline{\omega}_k^1$ that appear in the first term of equation [1.76], the stress $\widehat{\sigma}_{ji}^{II}$ can suffer a loss of symmetry, with the asymmetry given by equation [1.73]:

$$e_{ijk}\widehat{\sigma}_{ij}^{II} = \sum_{c\in\partial C^B} m_k^c = -\sum_{c\in C^B} e_{ijk}f_i^c l_j^c \qquad [1.79]$$

If no contact moments $\mathbf{m}^c$ exist along the boundary $\partial\Omega^B$, the stress asymmetry is zero. In this case, the strong equations of $\widehat{\sigma}^{II}$ and $\widehat{\mu}^{II}$ are decoupled (see equation [1.57]).

A third form of the internal virtual work is augmented by the higher-order displacement gradient $\delta\overline{u}_{ijk}$:

$$\delta W^{\text{Int,III}} = \widehat{\sigma}_{ji}^{III}(\delta\overline{u}_{ij}^2 - e_{ijk}\delta\overline{\omega}_k^1) + \widehat{\sigma}_{jki}^{III}(\delta\overline{u}_{ijk}^3 - e_{ij\ell}\delta\overline{\omega}_{\ell k}^2) + \widehat{\mu}_{ji}^{III}\delta\overline{\omega}_{ij}^2 \qquad [1.80]$$

in which the higher order-stress $\hat{\sigma}_{jki}^{III}$ is conjugate with the difference of the higher-order strain $\delta\overline{u}_{ijk}^{3}$ and the skew-symmetric part of the micro-rotation gradient $\delta\overline{\omega}_{\ell k}^{2}$ (the micro-curvature). The strong form of equilibrium for this generalized continuum is given by:

$$\left.\begin{array}{l} \sigma_{ji,j} + \sigma_{(jk)i,jk} + b_i = 0 \\ \mu_{ji,j} + e_{ijk}\sigma_{jk} - e_{ijk}\sigma_{j\ell k,\ell} + c_i = 0 \end{array}\right\} \text{ in } \Omega^{\mathcal{B}} \qquad [1.81]$$

$$\left.\begin{array}{l} \sigma_{ji}n_j - \sigma_{jki,j}n_k = t_i \\ \mu_{ji}n_j + e_{ijk}\sigma_{\ell k j}n_\ell = z_i \\ \sigma_{kji}n_j = \psi_{ij} \end{array}\right\} \text{ on } \partial\Omega^{\mathcal{B}} \qquad [1.82]$$

where $\mathbf{n}$ is the outward normal of surface $\partial\Omega^{\mathcal{B}}$, and ψ is a higher order traction.

A comparison of terms in equations [1.68] and [1.80] leads to the following expressions for the third set of macro-stresses:

$$\hat{\sigma}_{ji}^{III} = \frac{1}{V^{\mathcal{B}}} \sum_{c\in\partial C^{\mathcal{B}}} f_i^c x_j^c = \frac{1}{V^{\mathcal{B}}} \sum_{c\in C^{\mathcal{B}}} f_i^c l_j^c \qquad [1.83]$$

$$\hat{\mu}_{ji}^{III} = \frac{1}{V^{\mathcal{B}}} \sum_{c\in\partial C^{\mathcal{B}}} (m_i^c x_j^c - \frac{1}{2} e_{ik\ell} f_k^c x_\ell^c x_j^c)$$

$$= \frac{1}{V^{\mathcal{B}}} \sum_{c\in C^{\mathcal{B}}} \left[m_i^c l_j^c - e_{ik\ell} f_k^c \left(\frac{1}{2} J_{\ell j}^c - x_\ell^c l_j^c \right) \right] \qquad [1.84]$$

$$\hat{\sigma}_{jki}^{III} = \frac{1}{2V^{\mathcal{B}}} \sum_{c\in\partial C^{\mathcal{B}}} f_i^c x_j^c x_k^c = \frac{1}{2V^{\mathcal{B}}} \sum_{c\in C^{\mathcal{B}}} f_i^c J_{jk}^c \qquad [1.85]$$

Again, the appropriate macro-stress – type "I", "II" or "III" – depends on the particular continuum setting in which the discrete, granular medium is placed.

1.4. Granular deformation

Movement, settlement and flow are of primary concern in geotechnical and geologic situations. As with stress, the fields of movement and strain in a granular material are highly non-uniform, particularly at the scale of individual particles and particle clusters. Deformation within the nearly rigid grains is largely confined to the immediate vicinity of the contacts, and at small strains, these localized contact deformations are principally responsible for the bulk stiffness. At larger strains and during plastic flow, however, the rearrangement of particles accounts for most bulk deformation. Such particle rearrangements produce deformation of the void space rather than changes in the shapes or sizes of the particles themselves.

Movement and deformation are presented as fairly straightforward quantities in the mechanics literature, but usually in a continuum setting rather than the discontinuum context of granular materials. When the movements of individual material points are tracked, whether in a continuum or a discontinuum, it is essential to designate a reference condition, so that particle movements can be traced to this condition. Measures of strain in a continuum are based on the deformation gradient or the displacement gradient relative to the reference configuration. If $\mathbf{X}$ is the field of material points in the reference configuration (usually the initial condition) of a continuous region, and $\mathbf{x}$ is the corresponding field of the material points' positions in the deformed configuration, the deformation gradient $\mathbf{F}$ is the tensor of partial derivatives:

$$F_{ij} = \frac{\partial x_i}{\partial X_j} = \partial_j x_i \quad \text{or} \quad x_{i,j} \tag{1.86}$$

and sometimes written as $x_{i,J}$ to emphasize that the gradient is with respect to the reference system X_j (see [MAL 69]). We can also compute the displacements $\mathbf{u}$ of material points (e.g. the particles' reference points χ) in the current configuration from those in the reference configuration: $\mathbf{u} = \mathbf{x} - \mathbf{X}$. This displacement field is the basis of the displacement gradient $\mathbf{U}$:

$$U_{ij} = \frac{u_i}{X_j} = F_{ij} - \delta_{ij} = \partial_j u_i \quad \text{or} \quad u_{i,j} \tag{1.87}$$

where δ_{ij} is the Kronecker tensor (the identity matrix in Cartesian frames). The Green strain tensor $\mathbf{E}$ is defined in terms of the $\mathbf{F}$ or $\mathbf{U}$ tensors and is a symmetric and large-strain measure of deformation:

$$E_{ij} = \frac{1}{2}\left(\mathbf{F}^{\mathrm{T}}\mathbf{F} - \mathbf{I}\right) = \frac{1}{2}(F_{ki}F_{kj} - \delta_{ij}) = \frac{1}{2}(U_{ij} + U_{ji} + U_{ki}U_{kj}) \tag{1.88}$$

The change in length of an infinitesimal segment $d\mathbf{X}$ between the deformed and reference configurations is derived from the Green strain,

$$(ds)^2 - (dS)^2 = 2dX_i E_{ij} dX_j \qquad [1.89]$$

where ds and dS are the differential lengths in the deformed and reference systems. Although a small-strain approximation of $\mathbf{E}$, presented below, is often used in the analysis and reporting of results, large-strain measures such as $\mathbf{F}$, $\mathbf{U}$ and $\mathbf{E}$ are essential in numerical particle simulations, since the relative movements of particles at their contacts (often computed as an overlap) can be many orders of magnitude smaller than the particle size and yet further orders of magnitude smaller than the assembly (or specimen) size. Moreover, large fluctuations of particle movements from an affine condition of uniform strain are common in granular materials (section 4.2), such that a small-strain approximation of the local strains can lead to quite large errors in numerical computations. That being said, the symmetric small-strain approximation of strain, ε, is:

$$\varepsilon_{ij} = \frac{1}{2}(U_{ij} + U_{ji}) \qquad [1.90]$$

and the corresponding rotation is:

$$w_{ij} = \frac{1}{2}(U_{ij} - U_{ji}) \quad \text{or} \quad \omega_i = -\frac{1}{2}e_{ijk}U_{jk} \qquad [1.91]$$

expressed as the second-order tensor $\mathbf{w}$ or as the rotation vector ω.

The small-strain rotation in equation [1.91] is not appropriate when large rotations have taken place relative to the original configuration $\mathbf{X}$, since an additive decomposition of $\mathbf{F}$ into symmetric and skew symmetric parts does not represent a pure deformation and a pure rotation. For this purpose, the relevant rotation is given by the multiplicative decomposition:

$$F_{ij} = R_{ik}A_{kj} = B_{ik}R_{kj} \qquad [1.92]$$

where $\mathbf{R}$ is an orthonormal pure rotation (with $R_{ik}R_{jk} = R_{ki}R_{kj} = \delta_{ij}$), and $\mathbf{A}$ and $\mathbf{B}$ are the symmetric right stretch tensor and left stretch tensor.

These strain expressions are based on the displacements $\mathbf{u}$ of material points from their positions in the reference configuration. Similar expressions apply to the *velocities* **v** of material points and are usually presented as the velocity gradient tensor **L**. Unlike **F**, the velocity gradient **L** is referenced to the current configuration (the "**x** system"), and derivatives of the current velocity are with respect to position in the current configuration: $\partial v_i / \partial x_j$. If the movement information is instead referenced to the original configuration **X**, then the velocity gradient is easily computed from the deformation gradient **F** (relative to an original, reference configuration) and the rate at which the components of **F** are currently changing:

$$\mathbf{L} = \dot{\mathbf{F}}\mathbf{F}^{-1} \quad \text{or} \quad L_{ij} = \frac{\partial v_i}{\partial x_j} = \dot{F}_{ik}F_{kj}^{-1} \qquad [1.93]$$

The rate of deformation **D** and the spin **W** are defined as the symmetric and skew-symmetric parts of **L**:

$$\mathbf{D} = \frac{1}{2}\left(\mathbf{L} + \mathbf{L}^{\mathrm{T}}\right) \quad \text{and} \quad \mathbf{W} = \frac{1}{2}\left(\mathbf{L} - \mathbf{L}^{\mathrm{T}}\right) \qquad [1.94]$$

and the vorticity vector ω' is:

$$\omega'_i = -\frac{1}{2}e_{ijk}W_{jk} \qquad [1.95]$$

The definitions that have been presented thus far are straightforward when applied to a continuous medium in which the displacements are continuous and differentiable. The definitions are less equivocal for a granular material in which sliding between neighboring particles produces a discontinuous displacement field, in which the independent rotations of neighboring particles produce a discontinuous rotation field and in which deformations are abruptly different within adjoining solid and void regions. For this reason, several definitions of strain have been proposed for granular media. The most appropriate definition will usually depend on the purpose to which the strain is being applied, which includes the following possibilities:

– to interpret the results of experiments and simulations in which the particle movements are fully accessible. By computing local strains, such micro-scale studies can discern localized deformation zones, such as shear bands, the deformation patterns within these regions, the spatial distributions of local strains, possible evidence of material instability at small scales, and the local origin, development and manifestation of elasticity, plasticity and failure within a granular material;

– to build constitutive models based on the micro-mechanics of particle interactions. This "change of scale" problem requires a means of upscaling the micro-scale particle movements to a continuum measure of macro-scale strain;

– to implement multi-scale models for analyzing large geotechnical problems. In the FEM[2] approach to heterogeneous materials, the material behavior of a continuum (usually a finite element) model is informed by a micro-scale analysis of the material, and the discrete particle-scale data are homogenized to estimate continuum measures of stress and strain;

– to explore and analyze material behaviors in the context of advanced material models: non-local material models, micro-polar and micro-morphic models, and models in which the material response depends on the gradients of local, micro-scale strain and rotation.

Other uses of particle-level strain estimates will likely arise as analytic and computational capabilities increase, allowing the analysis of larger, more complex regions, while accounting for material deformation and behavior at small scales.

Method	Section	Tessellated region?	Contact or particle movements	2D only?
Simplex [BAG 96]	1.4.1.1	Y	C	N
Polygon [KUH 99]	1.4.1.2	Y	C	Y
Polygon [KRU 96]	1.4.1.2	Y	C	Y
Projection	1.4.2.1	N	P	N
Mesh-free [OSU 03]	1.4.2.2	N	P	N
Grid [WAN 07]	1.4.2.3	N	P	N
Contacts [LIA 97]	1.4.3.1	N	C	N
Contacts [KUH 04a]	1.4.3.2	Y/N	C	Y
Y = yes; N = no; C = contact movements; P = particle movements				

Table 1.3. Methods for computing or estimating strain in granular media

Table 1.3 lists several methods of estimating granular strains from particle-scale data. The table categorizes these methods in several respects:

1) whether the method requires tessellation of a larger region into small sub-regions within which the local strain is derived. Some methods do not require such tessellation and only require information on the locations and

movements of individual particles or on the relative movements of particle pairs;

2) whether the method applies to three-dimensional media or is exclusively intended for two-dimensional materials;

3) whether the method operates on the relative movements between particles (perhaps at their contacts) or upon the field of particle movements.

As the representative strain of a granular material is usually derived from the movements of nearly rigid grains, rather than from a continuous deformation field, the strain measures given below will be written with an overline: for example, $\overline{\mathbf{U}}$ will represent the discrete complement of the continuum deformation $\mathbf{U}$.

1.4.1. *Strain in tessellated regions*

These methods require a larger region to be divided into sub-regions of simplexes (triangles or tetrahedra), polygons or polyhedra, in which the vertices are located at particle reference points χ^p, perhaps at the particles' centers (Figures 1.2 and 1.4). An obvious measure of a representative deformation within a region $\Omega^{\mathcal{B}}$ is the volume-average displacement gradient within the region, expressed as a volume or surface integral,

$$\overline{U}_{ij}^{\mathcal{B}} = \frac{1}{V^{\mathcal{B}}} \int_{\Omega^{\mathcal{B}}} U_{ij}(\mathbf{x})\, dV = \frac{1}{V^{\mathcal{B}}} \int_{\Omega^{\mathcal{B}}} u_{i,j}\, dV \qquad [1.96]$$

$$= \frac{1}{V^{\mathcal{B}}} \int_{\partial\Omega^{\mathcal{B}}} u_i n_j\, dS \qquad [1.97]$$

where $\mathbf{n}$ is the outward normal of the region's surface $\partial\Omega^{\mathcal{B}}$. Although equation [1.97] is valid for any closed region, its application requires knowledge of the displacement $\mathbf{u}(\mathbf{x})$ along the boundary, rather than the relative movements between interior particles, and it assumes that a differentiable field of movements can be extended to the boundary. Most of the deformation measures detailed in this chapter involve the computation of strain from the particles' relative movements, either between particles that are touching or among the "virtual" contacts between particles that are nearby but are not touching.

1.4.1.1. *Strain in simplex regions*

We can approximate the movements of N points in a space of dimension D as conforming to the series:

$$u_i^p \approx A_i^{\bar{1}} + A_{ij}^{\bar{2}} X_j^p + \frac{1}{2} A_{ijk}^{\bar{3}} X_j^p X_k^p + \ldots \quad p = 1, 2, \ldots, N \quad [1.98]$$

in which $\mathbf{X}^p$ and $\mathbf{u}^p$ are components of the reference position and the displacement of the reference point χ^p assigned to particle p, and the overline numerals are array indicators, rather than exponents. If truncated at the second term, the approximation is an affine transformation, with $\mathbf{A}^{\bar{1}}$ as the rigid movement and $\mathbf{A}^{\bar{2}}$ corresponding to the displacement gradient $\mathbf{U} = \mathbf{F} - \mathbf{I}$ of equation [1.87]. When the number of points N is only $D + 1$, the system of equations [1.98] is neither over-constrained nor under-constrained, provided that the points $\mathbf{X}^p$ are neither coplanar nor collinear. This condition applies to simplexes of non-zero volume (triangles in 2D, tetrahedra in 3D). Bagi's method [BAG 96] is based on the spatial discretization of a 2D (or 3D) domain, by partitioning (tessellating) the domain into a covering of triangles (or tetrahedra) and computing the uniform affine deformation $\mathbf{A}^{\bar{2}}$ within each simplex (see Figure 1.2(c)).

Bagi [BAG 96] derived an expression for the average strain within a spatial region that has been tessellated into simplex sub-domains, as with the Delaunay triangulation. The vertices can be chosen as the reference points χ^p within the particles (Figure 1.4), in which case the edges represent either contacting pairs or "virtual contacts" between nearby (but non-touching) pairs of particles. Unlike equation [1.98], Bagi's strain is expressed in terms of the *relative* movements between the particles along the edges rather than their absolute movements $\mathbf{u}^p$. A single triangle (tetrahedron) has 3 (4) vertices and 3 (6) edges. Vector $\mathbf{b}^p$ is the outward-directed normal vector of the edge (face) that lies opposite the vertex p, and its magnitude $|\mathbf{b}^p|$ is equal to the length of the edge (area of the face). A corresponding vector $\mathbf{a}^p = -(1/D)\mathbf{b}^p$ has a direction opposite that of $\mathbf{b}^p$ and has a magnitude $1/D$ times that of $\mathbf{b}^p$. An initial vertex p and a terminal vertex q are assigned to each edge c of the simplex, and the relative displacement Δ^c and the relative directors α^c and β^c of each edge are defined as:

$$\Delta_i^c = u_i^q - u_i^p, \quad \alpha_i^c = a_i^q - a_i^p, \quad \beta_i^c = b_i^q - b_i^p \quad [1.99]$$

By assuming that displacement $\mathbf{u}(\mathbf{X})$ varies linearly along the boundary of the simplex region, the gradient $\mathbf{A}^{\tilde{2}}$ within a single simplex L is equal to its average displacement gradient $\overline{\mathbf{U}}^L$ and is computed from the sum of dyads $\mathbf{\Delta} \otimes \boldsymbol{\alpha}$ or $\mathbf{\Delta} \otimes \boldsymbol{\beta}$ of all edges:

$$U_{ij}^L = \frac{1}{D+1} \frac{1}{V^L} \sum_c \Delta_i^c \alpha_j^c = -\frac{1}{D(D+1)} \frac{1}{V^L} \sum_c \Delta_i^c \beta_j^c \qquad [1.100]$$

where V^L is the volume of simplex Ω^L. For triangles (2D), the sum is of the three edges; for tetrahedra (3D), the sum is of six edges.

We can use equation [1.100] to examine the local strains within a granular region that has been tessellated into a covering of simplexes. The average strain $\overline{\mathbf{U}}^{\mathcal{B}}$ within the tessellated region $\Omega^{\mathcal{B}}$ is found as the volume average of the simplex strains:

$$\overline{U}_{ij}^{\mathcal{B}} = \frac{1}{V^{\mathcal{B}}} \sum_{L \in \Omega^{\mathcal{B}}} V^L U_{ij}^L \qquad [1.101]$$

where $V^{\mathcal{B}}$ is the volume of region $\Omega^{\mathcal{B}}$. Bagi simplified this expression by assigning a *complementary area vector* $\mathbf{d}^c$ to each edge c within region $\Omega^{\mathcal{B}}$: "collect now all those [simplexes] that contain this edge [c]. Assume that altogether T cells were found". That is, the edge c is shared by T different faces and T different simplexes, and each of these faces is a triangle (2D) or the triangular face of a tetrahedron (3D). The complementary area vector is defined as:

$$d_i^c = \frac{1}{D+1} \sum_{t=1}^{T} \left(a_i^{t,q} - a_i^{t,p} \right) \qquad [1.102]$$

where $\mathbf{a}^{t,p}$ and $\mathbf{a}^{t,q}$ are the $\mathbf{a}$-vectors corresponding to the p and q vertices of the t-simplex that contains the edge c (directed from vertex p to vertex q). With these $\mathbf{d}$-vectors, the volume-average strain among the simplexes that comprise $\Omega^{\mathcal{B}}$ is:

$$\overline{U}_{ij} = \frac{1}{V^{\mathcal{B}}} \sum_{c \in \Omega^{\mathcal{B}}} \Delta_i^c d_j^c \qquad [1.103]$$

This expression for strain has a similar form as that of the stress in equation [1.53]: the average strain is a sum of the dyads $\mathbf{\Delta}^c \otimes \mathbf{d}^c$ that are associated with the edges, representing both contacts and virtual contacts, of a simplex covering of region $\Omega^{\mathcal{B}}$.

1.4.1.2. Strain in polygon regions

The author's definition of strain is a variation of the Bagi [BAG 96] definition but is intended exclusively for two-dimensional regions $\Omega^{\mathcal{B}}$ [KUH 99]. The region is tessellated into polygonal sub-regions that are defined by the contacts between particles (Figure 1.3(a)). This geometric tessellation, the *particle graph*, was suggested by Satake [SAT 92] to represent the internal structure of a two-dimensional granular medium and is the geometric expression of a planar graph of vertices (particle reference points χ^p), edges (connections of neighboring particles through contacts) and faces (polygonal regions containing both voids and solids). In this representation, some unconnected particles can lie within the void polygons, and these *rattlers* are ignored in computing the average deformation. The edges and vertices of a polygon are numbered in a counterclockwise manner, beginning with zero and ending with $m^L - 1$, where m^L is the valence of the polygon Ω^L (Figure 1.8). The edges are viewed as branch vectors $\mathbf{l}^k$, joined head-to-tail around the void polygon (Figure 1.8(a)). A vector $\mathbf{b}^k$ is associated with the kth edge between particles p and q, having a length equal to that of the branch vector between the two particles, $|\mathbf{l}^k|$, and directed outward from the polygon (Figure 1.8(b)). The average displacement gradient $\overline{\mathbf{U}}^L$ within polygon Ω^L is the matrix product of an $m \times m$ matrix $[\mathbf{Q}^m]_{3 \times 3}$ and the dyads $\mathbf{\Delta}^{k_1} \otimes \mathbf{b}^{k_2}$ taken counterclockwise around the polygon,

$$\overline{U}^L_{ij} = \frac{1}{6A^L} \sum_{k_1}^{m-1} \sum_{k_2}^{m-1} Q^m_{k_1,k_2} \Delta^{k_1}_i b^{k_2}_j. \tag{1.104}$$

where A^L is the area of polygon Ω^L, and vector $\mathbf{\Delta}^k$ is the relative displacement of two adjacent vertices: $\Delta^k_i = u^{k+1}_i - u^k_i$, with $k + 1$ taken in modulo m: $k + 1 = \mathrm{mod}(k + 1, m)$. The 3×3 matrix for a triangle is:

$$[\mathbf{Q}]^3 = \begin{bmatrix} 0 & 1 & -1 \\ -1 & 0 & 1 \\ 1 & -1 & 0 \end{bmatrix}. \tag{1.105}$$

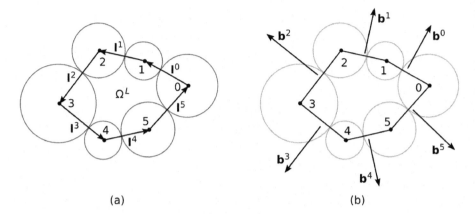

Figure 1.8. *Void polygon Ω^L with $m = 6$ edges, as used in computing the average strain within a two-dimensional region: (a) branch vectors $\mathbf{l}^k$ and (b) outward normal vectors $\mathbf{b}^k$, with lengths $|\mathbf{b}^k| = |\mathbf{l}^k|$, and $k = 0, 1, \ldots, m - 1$ [KUH 99]*

In general, matrix $[\mathbf{Q}^m]$ has a skew-symmetric and circulant form, and a recursive expression for the $[\mathbf{Q}^m]$ matrix of a general m-polygon was derived in [KUH 97]. Table 1.4 gives the first row of the $[\mathbf{Q}^m]$ matrices for triangles through hexagons, with

$$Q^m_{k_1,k_2} = \begin{cases} 0 & k_1 = k_2 \\ 3 - \dfrac{6}{m}\mathrm{mod}(k_2 - k_1, m) & k_1 \neq k_2 \end{cases} \qquad [1.106]$$

Valence	$k_2 - k_1$ (mod m)					
m	0	1	2	3	4	5
3	0	$\frac{3}{3}$	$-\frac{3}{3}$			
4	0	$\frac{6}{4}$	0	$-\frac{6}{4}$		
5	0	$\frac{9}{5}$	$\frac{3}{5}$	$-\frac{3}{5}$	$-\frac{9}{5}$	
6	0	$\frac{12}{6}$	$\frac{6}{6}$	0	$-\frac{6}{6}$	$-\frac{12}{6}$

Table 1.4. *First row of matrix $Q^m_{k_1,k_2}$*

As this form of $[\mathbf{Q}^m]$ is circulant, the remaining rows are given as $Q^m_{pq} = Q^m_{p-1,q-1}$ with the subscripts computed modulo m: each successive row is shifted by one column position. As in equation [1.101], the average displacement gradient within an entire region $\overline{\mathbf{U}}^{\mathcal{B}}$ is the area-averaged sum of the polygon gradients $\overline{\mathbf{U}}^{L}$.

Similar to the author's definition, the strain definition of Kruyt and Rothenburg [KRU 96] is intended for a two-dimensional region that is tessellated into polygonal sub-regions. They recognized that the boundary unit normal $\mathbf{n}$ in the integral of equation [1.96], when rotated 90°, coincides with a boundary's counterclockwise tangent vector $\mathbf{t}$, such that $n_i = R_{ij}t_j$, where rotation matrix R_{ij} is given by:

$$\mathbf{R} = \begin{bmatrix} 0 & 1 \\ -1 & 0 \end{bmatrix} \qquad\qquad [1.107]$$

As $t_i = dx_i/ds$ along the boundary arc ds, the average displacement gradient within polygon L, in accordance with equation [1.97], is:

$$\overline{U}_{ij}^{L} = \frac{1}{V^L} \int_{\partial\Omega^L} R_{jk} u_i \frac{dx_k}{ds}\, ds = -\frac{1}{V^L} \int_{\partial\Omega^L} R_{jk} \frac{du_i}{ds} x_k\, ds \qquad [1.108]$$

If the derivative du_i/ds in the second integral is constant along the straight edge of a polygon (i.e. an affine displacement along the edge), the integral along the edge is equal to the difference Δ in the displacements of the edges' two vertices (i.e. $\Delta = \mathbf{u}^q - \mathbf{u}^p$), and the integral is reduced to a sum of contributions from each of the polygon's m^L edges:

$$\overline{U}_{ij}^{L} = -\frac{1}{V^L} R_{jk} \sum_{\ell=1}^{m^L} \Delta_i^{\ell} x_k^{\ell} \qquad\qquad [1.109]$$

where Δ^{ℓ} is the relative displacement of the two vertices (i.e. the particle reference points χ^p) of branch vector $\mathbf{l}^{\ell}$ as we move counterclockwise around the polygon. Position $\mathbf{x}^{\ell}$ is the midpoint of this branch vector.

The average deformation of equation [1.108] is similar to the equation for average stress in a particle, equation [1.46], and noting this and other similarities, Kruyt and Rothenburg [KRU 96] derived an expression for the average strain within a region. Their expression uses only the relative movements of neighboring particles and the relative positions of the neighboring void polygons. The compatibility of relative movements Δ^{ℓ} around a single polygon L is:

$$\sum_{\ell=1}^{m^L} \Delta_i^{\ell} = 0 \qquad [1.110]$$

similar to the equilibrium condition of equation [1.45]. Finally, as with the action–reaction relation $\mathbf{f}^{pq} = -\mathbf{f}^{qp}$ of contact forces between particles that share a common contact, the relative displacements Δ^{ℓ} along the common edge of two neighboring void polygons are equal in magnitude but of opposite (counterclockwise) directions. The average displacement gradient within the larger region $\Omega^{\mathcal{B}}$ is the sum:

$$\overline{U}_{ij}^{\mathcal{B}} = \frac{1}{V^{\mathcal{B}}} \sum_{c \in C^{\mathcal{B}}} \Delta_i^c h_j^c \qquad [1.111]$$

where $C^{\mathcal{B}}$ is the set of edges (contacts) in the region. Each edge is assigned a polarity (say, from vertex p to vertex q), such that $\Delta^c = \mathbf{u}^q - \mathbf{u}^p$, and the polygon vector $\mathbf{h}^c$ is a counterclockwise rotation of the polygon vector $\mathbf{g}^c$ that connects the center of the polygon to the left of vertex p to the center of the polygon on its right (see Figure 1.9). Specifically,

$$h_i^c = -R_{ij} g_j^c \qquad [1.112]$$

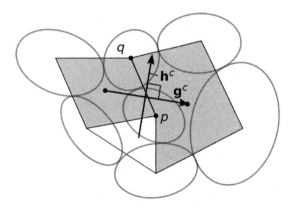

Figure 1.9. *Two void polygons with edge $\mathbf{l}^{pq} = \mathbf{l}^c$ between particles p and q, vector $\mathbf{g}^c$ between the polygon centers and its orthogonal vector $\mathbf{h}^c$*

Just as the branch vector $\mathbf{l}^c$ is the (directed) vector between the centers of two neighboring particles, vector $\mathbf{g}^c$ is the directed vector between the centers of two neighboring void polygons and void vector $\mathbf{h}^c$ is its orthogonal complement. A comparison of equations [1.51] and [1.111] reveals that they are dual relations for stress and strain.

1.4.2. *Strain without spatial tessellation*

The strain definitions of the previous sections require the spatial discretization (tessellation) of a granular region into sub-regions that each contain a single particle or a small cluster of particles. Tessellating a region into such well-defined non-overlapping sub-regions can be particularly difficult with experimental data or with simulations of particles having irregular shapes, and the tessellation can also be difficult with assemblies of spheres having a wide range of diameters. Other methods avoid such partitioning by computing representative local strains in possibly overlapping sub-regions, perhaps defined with a grid covering the full region. A representative strain is computed in each sub-region from the movements of particles within the sub-region, although this region does not need to be associated with any particular particle. As strain within a region is based on the movements of material points, the tracking of multiple points within individual particles can even capture the rotations as well as the translations of these particles. While strain definitions in the previous sections are based on affine deformations (i.e. movement fields restricted to the two terms $A_i^{\bar{1}}$ and $A_{ij}^{\bar{2}}$ in equation [1.98]), this restriction can be removed for the methods of the current section.

1.4.2.1. *Projection method*

An obvious approach for estimating strain within a sub-region is to compute the terms of $A_{ij}^{\bar{2}}$ in equation [1.98] that achieve a best-fit to the measured displacements of points (attached to particles) within the region. If the displacements of only three points are tracked in a 2D assembly (four points in 3D), a unique linear fit of equation [1.98] can be found, provided that the points are not collinear (coplanar in 3D). With more points, equation [1.98] is over-constrained, but we can readily compute the projection of the movement data onto a sub-space of affine deformations [STR 11]. In 3D, the elements of the mean displacement $A_i^{\bar{1}}$ and the displacement gradient

$A_{ij}^{\bar{2}}$ are arranged as a vector of 12 values $[\mathbf{A}] = [A_1^{\bar{1}}, A_2^{\bar{1}}, A_3^{\bar{1}}, A_{11}^{\bar{2}}, \dots, A_{33}^{\bar{2}}]^{\mathrm{T}}$, which include the rigid translation $\mathbf{A}^{\bar{1}}$ terms and the affine deformation $\mathbf{A}^{\bar{2}}$ terms. These 12 values are approximated with the product of a pseudo-inverse matrix $[\mathbf{G}]^+$ multiplied by the $3N$ displacement components $\mathbf{u}^p$ of the N tracked reference points χ^p:

$$[\mathbf{A}] = [\mathbf{G}]^+[\mathbf{u}] \tag{1.113}$$

where the Moore–Penrose pseudo-inverse matrix is computed as:

$$[\mathbf{G}]^+ = \left([\mathbf{G}]^{\mathrm{T}}[\mathbf{G}]\right)^{-1}[\mathbf{G}]^{\mathrm{T}} \tag{1.114}$$

and $[\mathbf{G}]$ are $[\mathbf{u}]$ are arrangements of the N positions $\mathbf{x}^p$ and the N movements $\mathbf{u}^p$ of the material (reference) points:

$$[\mathbf{G}] = \begin{bmatrix} \mathbf{I}_{\bar{3}} & x_1^1\mathbf{I}_{\bar{3}} & x_2^1\mathbf{I}_{\bar{3}} & x_3^1\mathbf{I}_{\bar{3}} \\ \mathbf{I}_{\bar{3}} & x_1^2\mathbf{I}_{\bar{3}} & x_2^2\mathbf{I}_{\bar{3}} & x_3^2\mathbf{I}_{\bar{3}} \\ \vdots & \vdots & \vdots & \vdots \\ \mathbf{I}_{\bar{3}} & x_1^N\mathbf{I}_{\bar{3}} & x_2^N\mathbf{I}_{\bar{3}} & x_3^N\mathbf{I}_{\bar{3}} \end{bmatrix}_{3N\times12} \qquad [\mathbf{u}] = \begin{bmatrix} [u_1^1, u_2^1, u_3^1]^{\mathrm{T}} \\ [u_1^2, u_2^2, u_3^2]^{\mathrm{T}} \\ \vdots \\ [u_1^N, u_2^N, u_3^N]^{\mathrm{T}} \end{bmatrix}_{3N\times1} \tag{1.115}$$

Matrix $\mathbf{I}_{\bar{3}}$ is the 3×3 identity matrix. This best-fit approach can be applied to regions of a few particles or to entire granular assemblies. The method can be used for two- and three-dimensional assemblies [GUO 14] and is part of the PFC3D software for DEM simulation [ITA 14]. The method can be extended to non-uniform strain fields – fields defined by a combination of strain and the gradients of strain – by using a quadratic approximation of the displacement field that includes the third-order tensor $A_{ijk}^{\bar{3}}$ in equation [1.98] [ALS 06].

1.4.2.2. Mesh-free strain method

Drawing upon mesh-free homogenization methods used in finite element analyses, O'Sullivan et $al.$ [OSU 03] proposed a method for interpolating strains within two- or three-dimensional granular media. Representative strains are computed for a small region that surrounds a single point (i.e. the point's support). Although the points at which the strains are computed can be uniformly spaced points on a rectangular grid, the use of a grid is not necessary, as the points can be assigned to irregular locations within a region, and can even be placed at the centers of particles. The displacement associated with a point $\mathbf{x}$ is an interpolated, averaged value that is computed from the known displacements $\mathbf{u}^p$ of surrounding material points p at positions $\mathbf{x}^p$ that lie within the support $\Omega^{\mathbf{x}}$ of point $\mathbf{x}$. These displacements of nearby tracked points can be the movements of particle centers or the movements of a select number of points within a particle or group of particles (e.g. see [LI 02a]):

$$u_i(\mathbf{x}) \approx \sum_{p \in \Omega^{\mathbf{x}}} W(\mathbf{x}^p - \mathbf{x}) u_i^p V^p \qquad [1.116]$$

where $p \in \Omega^{\mathbf{x}}$ are the tracked points near $\mathbf{x}$ and within its support $\Omega^{\mathbf{x}}$, W is a scalar smoothing kernel function, and $\mathbf{x}^p$, $\mathbf{u}^p$ and V^p are the position, movement and volume (weight) associated with the tracked point p. The displacement gradient at $\mathbf{x}$, the gradient $\overline{U}(\mathbf{x})$, is computed as:

$$\overline{U}_{ij}(\mathbf{x}) \approx \sum_{p \in \Omega^{\mathbf{x}}} \left(\frac{\partial}{\partial x_j} W(\mathbf{x}^p - \mathbf{x}) \right) u_i^p V^p \qquad [1.117]$$

Applying equation [1.117] to find the strain at $\mathbf{x}$ requires identifying all tracked material points and conducting a near-neighbor search to find those tracked points that lie within the support of the strain point $\mathbf{x}$. Although a Gaussian kernel is a natural choice for the smoothing function, a function with compact support is more computationally efficient. O'Sullivan *et al.* [OSU 03] suggest a composite product of cubic splines:

$$W(\mathbf{y}) = \begin{cases} \dfrac{1}{\rho^2} \phi\left(\dfrac{y_1}{\rho}\right) \phi\left(\dfrac{y_2}{\rho}\right) & \text{2D} \\[3mm] \dfrac{1}{\rho^3} \phi\left(\dfrac{y_1}{\rho}\right) \phi\left(\dfrac{y_2}{\rho}\right) \phi\left(\dfrac{y_3}{\rho}\right) & \text{3D} \end{cases} \qquad [1.118]$$

where ρ is a parameter that defines the quarter-width of the interpolation region (support) around point $\mathbf{x}$:

$$\phi(z) = \begin{cases} \frac{1}{6}(z+2)^3 & z \in [-2,-1] \\[1mm] \frac{2}{3} - z^2(1+z/2) & z \in [-1,0] \\[1mm] \frac{2}{3} - z^2(1-z/2) & z \in [0,1] \\[1mm] \frac{1}{6}(z-2)^3 & z \in [1,2] \\[1mm] 0 & z \notin [-2,2] \end{cases} \qquad [1.119]$$

The tracked points are weighted by values V^p, which can represent the volume associated with points p around the grid point, provided that the weights satisfy the normalizing condition:

$$\sum_{p \in \Omega^{\mathbf{x}}} W(\mathbf{x}^p - \mathbf{x})V^p = 1 \qquad [1.120]$$

O'Sullivan *et al.* [OSU 03] used the product of a physical volume associated with point p (computed from a Delaunay triangulation of all tracked points) and a corrector function that is required to assure that condition [1.120] is met. Other reasonable methods, not requiring a triangulation, can also be used to assign values to the V^p.

A width ρ must also be chosen for the interpolation region. Large widths will average the displacements of larger regions at the risk of "smearing" the localized strain in which we might be interested. Of course, widths smaller than the particle size will capture only the more erratic field of the discrete particle movements and rotations. O'Sullivan [OSU 03] used a ρ equal to the particle size, so that the width of the interpolating region was twice the particle size.

1.4.2.3. *Strain within a deforming grid*

Wang *et al.* [WAN 07] proposed a method in which a closely spaced rectangular grid covers a two- or three-dimensional granular region, with the grid spacing being on the order of the average particle size. The grid points (nodes) are displaced in accordance with the particle movements, as described below. Once deformed, the local strains can be computed from the nodal displacements, by using standard methods encountered in finite element analysis for computing element strains from nodal displacements (see [BAT 76]). The nodal displacements are found by assigning each grid point to its nearest particle, as determined by the distance vector $\mathbf{d}^g = \mathbf{x}^g - \mathbf{x}^p$ between a particle p and the grid point g. Wang *et al.* [WAN 07] used a special metric in assigning grid points to particles: the particle assigned to a given grid point minimizes the distance $|\mathbf{d}^g/r^p|$, where r^p is the particle's radius or size. Although each grid point is assigned to a single particle, a particle can be associated with multiple grid points. The grid points are then displaced in accordance with its particle's displacement and rotation:

$$d\mathbf{u}^g = d\mathbf{u}^p + d\theta^p \times \mathbf{d}^g \qquad [1.121]$$

and these nodal movements are then used to compute local strains within the grid.

1.4.3. *Bulk strain estimates for up-scaling analyses*

When deriving the bulk constitutive behavior of a granular material from the micro-scale interactions of its particles, it is natural to focus on the contacting particles and on the interactions of these particle pairs. Estimates of strain that are based solely on the contact movements and contact characteristics are particularly useful in such upscaling analyses. Upscaling from the particle movements to the bulk deformation requires a means of estimating bulk strain from the relative movements of particle pairs. The most naive upscaling methods ignore the topology of the contact network – the arrangement of particles into contacting pairs and the larger arrangement of these pairs into clusters and entire assemblies – and these methods estimate the strain based solely on the relative movements of two-particle systems. Of course, an exact strain cannot be extracted from such sparse information, but these estimates can serve their purpose in developing approximate constitutive models. Two such estimates are presented in this section.

1.4.3.1. *Up-scaling strains from relative movements*

The method of Liao *et al.* [LIA 97] is based on a best-fit approach and is applicable to two- and three-dimensional granular systems. Liao *et al.* [LIA 97] estimated the relative movements between the centers of two particles at a contact c as the product of the bulk strain $\overline{U}$ and the branch vector l^c that connects the two particles' reference points (Figure 1.4):

$$\Delta_i^c \approx \overline{U}_{ij} l_j^c \qquad [1.122]$$

where Δ^c are the relative translations of the particles' reference points (the movement $\Delta^c = \mathbf{u}^q - \mathbf{u}^p$, with $l^c = \mathbf{x}^q - \mathbf{x}^p$). This equation applies a uniform, affine displacement approximation to the movements, although the actual movements will likely differ from this approximation. When applied to all contacts in an assembly or region, $c \in \Omega^{\mathcal{B}}$, the equations [1.122] are an over-constrained system of linear equations. As such, the bulk deformation can be approximated as the projection of the actual movements onto an affine field. That is, rather than estimating the movements Δ^c from strain $\overline{U}$, the strain is estimated from the movements by multiplying by an appropriate pseudo-inverse matrix:

$$[\overline{U}] = [G]^+[\Delta] \tag{1.123}$$

In three dimensions, the system of equations [1.122] applies to the M contacts in an assembly as follows:

$$
\begin{bmatrix}
l_1^1 & l_2^1 & l_3^1 & 0 & 0 & 0 & 0 & 0 & 0 \\
0 & 0 & 0 & l_1^1 & l_2^1 & l_3^1 & 0 & 0 & 0 \\
0 & 0 & 0 & 0 & 0 & 0 & l_1^1 & l_2^1 & l_3^1 \\
 & \vdots & & & \vdots & & & \vdots & \\
l_1^c & l_2^c & l_3^c & 0 & 0 & 0 & 0 & 0 & 0 \\
0 & 0 & 0 & l_1^c & l_2^c & l_3^c & 0 & 0 & 0 \\
0 & 0 & 0 & 0 & 0 & 0 & l_1^c & l_2^c & l_3^c \\
 & \vdots & & & \vdots & & & \vdots & \\
l_1^M & l_2^M & l_3^M & 0 & 0 & 0 & 0 & 0 & 0 \\
0 & 0 & 0 & l_1^M & l_2^M & l_3^M & 0 & 0 & 0 \\
0 & 0 & 0 & 0 & 0 & 0 & l_1^M & l_2^M & l_3^M
\end{bmatrix}_{3M \times 9}
\begin{bmatrix}
\overline{U}_{11} \\
\overline{U}_{12} \\
\overline{U}_{13} \\
\overline{U}_{21} \\
\overline{U}_{22} \\
\overline{U}_{23} \\
\overline{U}_{31} \\
\overline{U}_{32} \\
\overline{U}_{33}
\end{bmatrix}
=
\begin{bmatrix}
\Delta_1^1 \\
\Delta_2^1 \\
\Delta_3^1 \\
\vdots \\
\Delta_1^c \\
\Delta_2^c \\
\Delta_3^c \\
\vdots \\
\Delta_1^M \\
\Delta_2^M \\
\Delta_3^M
\end{bmatrix}_{3M \times 1}
\tag{1.124}
$$

where $[G]$ appears as the matrix on the left. The product of its pseudo-inverse (as in equation [1.114]) and the column vector $[\Delta]$ gives the Liao *et al.* estimate of strain [LIA 97]:

$$\overline{U}_{ij} \approx \frac{1}{V} A_{jk} \sum_{c \in C^{\mathcal{B}}} \Delta_i^c l_k^c \tag{1.125}$$

where **A** is the symmetric tensor:

$$A_{jk} = \left(\frac{1}{V} \sum_{c \in C^{\mathcal{B}}} l_j^c l_k^c \right)^{-1} \tag{1.126}$$

such that $\overline{U}$ is estimated from the relative particle movements Δ^c. The author has found that the estimate in equation [1.125] is within 20% of the true strain for deviatoric deformation.

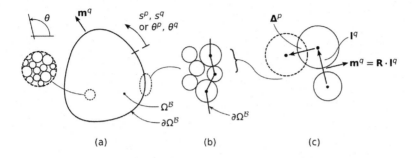

Figure 1.10. *Boundary chain of particles for finding the average velocity gradient within a two-dimensional region: (a) bounded assembly of particles with perimeter length* S *, (b) boundary chain of branch vectors and (c) two boundary particles*

1.4.3.2. *Up-scaling method for 2D regions*

The author has shown that the average deformation gradient $\overline{\mathbf{U}}$ within a two-dimensional region is exactly given by a double integral around its perimeter [KUH 04a]. Perimeter locations are parameterized by the arc distances s^p and s^q measured counterclockwise from a common, fixed point on the boundary (Figure 1.10(a)). The area-averaged velocity gradient is:

$$\overline{U}_{ij} = \frac{1}{A} \int_0^S \int_0^S \frac{du_i}{ds^p} Q(s^p, s^q) m_j^q \, ds^p ds^q \qquad [1.127]$$

where S is the perimeter length, $\mathbf{m}^q$ is the unit outward normal of the perimeter at position s^q, A is the enclosed area, $d\mathbf{u}/ds^p$ is the derivative vector of displacement $\mathbf{u}$ along the perimeter, as point s^p traverses this boundary and kernel Q is a discontinuous function $[0, S) \times [0, S)$ range $(-1/2, 1/2)$ given by:

$$Q(s^p, s^q) = \begin{cases} \dfrac{1}{2} - \dfrac{1}{S} \bmod(s^q - s^p, S), & s^q \neq s^p \\ 0, & s^q = s^p \end{cases} \qquad [1.128]$$

using the modulo mod( ) function. Application of integral [1.127] requires knowledge of the boundary normal $\mathbf{m}^q$ and the derivative $d\mathbf{u}/ds^p$ along the perimeter, which is related to the *relative motions* between pairs of boundary particles. The equation is a general and exact expression for the average spatial gradient within a closed two-dimensional region (bounded by a Jordan curve), provided that the boundary derivative $d\mathbf{u}/ds$ is integrable and consistent, such that $\int_0^S (du_i/ds)\,ds = 0$, and $\int_0^S m_j\,ds = 0$. The integral can account for the rotations of boundary particles and the abrupt displacements that occur at the boundary contact points between rigid particles [KUH 04a].

For an assembly of particles, a closed polygonal chain of M segments (branch vectors) is formed by the branch vectors between particles along the assembly's perimeter (Figure 1.10(b)). The perimeter (branch) segment q between two particles has an outward unit normal $\mathbf{m}^q$ that is perpendicular to the branch vector $\mathbf{l}^q$, such that $m_j^q = R_{jk}l_k^q/|\mathbf{l}^q|$, where $|\mathbf{l}^q|$ is the length of the branch vector and the rotation matrix R_{jk} is defined in equation [1.107].

In the most basic application of equation [1.127], bulk deformation is assumed to be entirely due to the relative motions between the centers of contacting particles, and each boundary branch vector $\mathbf{l}^p$ is assumed to be deformed in an affine manner, such that $d\mathbf{l}^p/ds$ is constant along the length $|\mathbf{l}^p|$ of its boundary branch. That is, we ignore particle rotations in this basic application and ignore the rigidity of the particles and the consequent concentration of relative movements at the contacts between particles. The derivative $d\mathbf{u}^p/ds$ equals $\boldsymbol{\Delta}^p/|\mathbf{l}^p|$, where $\boldsymbol{\Delta}^p$ is the relative displacement of the two particles that form branch p. As the derivatives and the boundary normals are assumed to be constant within each branch vector along an assembly's boundary, equation [1.127] can be simplified as the double sum:

$$\overline{U}_{ij} = \frac{1}{A} \sum_{p=1}^{M} \sum_{q=1}^{M} \Delta_i^p l_k^q R_{jk} Q(s^p, s^q) \qquad [1.129]$$

where M is the number of branch segments along the perimeter (with $p, q \in \partial C^{\mathcal{B}}$), and s^p and s^q are arc lengths from the reference point on the boundary to the centers of segments p and q (Figures 1.10(a) and (c)). This equation is the exact average strain within a region enclosed within a piecewise linear boundary and along which the derivative $d\mathbf{u}^p/ds$ is piecewise constant and integrable.

Although equation [1.129] is an exact expression of average strain in a two-dimensional region, an approximation is derived from the probability distribution of contact orientations and contact movements [KUH 16c]. If θ is the orientation angle of a branch vector and $p(\theta, \boldsymbol{\Delta})$ is the joint probability density of contact orientation and contact movement, then equation [1.129] can be cast in an approximate form as the double integral:

$$\overline{U}_{ij} \approx \frac{(M\bar{l})^2}{A} \int_0^{2\pi} \int_0^{2\pi} \frac{\Delta_i^p}{\bar{l}} \frac{l_k^q}{\bar{l}} R_{jk} Q^\theta(\theta^p, \theta^q) p(\theta, \Delta) \, d\theta^p d\theta^q \qquad [1.130]$$

where $\bar{l}$ is the average length of a branch vector, M is the number of perimeter contacts that encompass the two-dimensional region of area A, **R** is the rotation matrix of equation [1.107] and Q^θ is a discontinuous function of domain $[0, 2\pi) \times [0, 2\pi)$ and range $(-\frac{1}{2}, \frac{1}{2})$:

$$Q^\theta(\theta^p, \theta^q) = \begin{cases} \dfrac{1}{2} - \dfrac{1}{2\pi} \mathrm{mod}(\theta^q - \theta^p, 2\pi) & \theta^q \neq \theta^p \\ 0 & \theta^q = \theta^p \end{cases} \qquad [1.131]$$

Equation [1.130] is an approximation of the average strain, with errors of about 10% [KUH 16c]. These errors result from a number of sources, including the subtle correlations among the branch vector lengths and branch vector orientations, and correlations between contact movements and orientations that are not fully captured in the density $p(\theta, \Delta)$.

2

Contact Interaction

The previous chapter was concerned with bulk characteristics of granular materials: fabric, stress and strain. We now turn to the micro-scale and focus on the interactions of particle pairs. During bulk deformation, grains move and interact with their neighbors, largely through interactions at inter-particle contacts. Contact movements alter the inter-particle forces and result in changes of the bulk stress. The movements and forces between a single pair of particles are the subjects of this chapter.

2.1. Contact kinematics

A precise resolution of inter-particle movements is required when developing numerical models of particle systems or when interpreting the results of physical loading tests. Resolving the relative movements of particles is also necessary when developing, testing and applying analytical expressions of the contact force–movement relationship. These expressions can be quite complex, even for the simplest case of a monotonically advancing movement that only engages a single force mechanism, the subject of later sections. Moreover, when a granular material is deformed, multiple mechanisms – indentation, sliding, rolling, twisting, spinning, etc. – typically occur simultaneously among the particles at their contacts, so a precise definition of each mechanism is required.

When reduced to an elemental form, the kinematic problem is to fully describe the motions associated with a contact c that result from the movements of its two touching particles, p and q (Figures 1.4 and 2.1).

Reference points χ^p and χ^q are attached to the particles, and the particles' translations and rotations are tracked at these two points. The relative translation of p and q is the movement $d\mathbf{u}^{pq}$:

$$d\mathbf{u}^{pq} = d\mathbf{u}^q - d\mathbf{u}^p \qquad [2.1]$$

where $d\mathbf{u}^p$ and $d\mathbf{u}^q$, each with three Cartesian components, are the translations of the two reference points during the time increment dt. (Note that this use of small, incremental movements departs from the finite deformations of section 1.4.) Contact vectors $\mathbf{r}^{c,p}$ and $\mathbf{r}^{c,q}$ connect the reference points to the contact, which is located at an idealized point $\mathbf{x}^c$: that is, $\mathbf{r}^{c,p} = \mathbf{x}^c - \mathbf{x}^p$ and $\mathbf{r}^{c,q} = \mathbf{x}^c - \mathbf{x}^q$. The branch vector $\mathbf{l}^c$ is directed between the two reference points, $\mathbf{l}^c = \mathbf{x}^q - \mathbf{x}^p = \mathbf{r}^{c,p} - \mathbf{r}^{c,q}$.

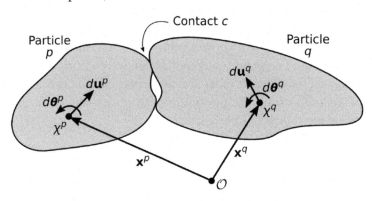

Figure 2.1. *Two contacting particles, p and q. Each particle can translate and rotate with incremental movements $d\mathbf{u}$ and $d\theta$ (also see Figure 1.4)*

The particles can also rotate by the increments $d\theta^p$ and $d\theta^q$ during dt. The four movement vectors, $d\mathbf{u}^p$, $d\mathbf{u}^q$, $d\theta^p$ and $d\theta^q$, are described by twelve scalar components, which form a 12-dimensional vector space of possible movements. For the moment, we assume that the particles are perfectly rigid but can inter-penetrate (overlap) in the vicinity of their contact c. The relative movement of two points near c, one attached to p and the other to q, is the "rigid" increment:

$$d\mathbf{u}^{c,\text{rigid}} = (d\mathbf{u}^q - d\mathbf{u}^p) + (d\theta^q \times \mathbf{r}^{c,q} - d\theta^p \times \mathbf{r}^{c,p}) \qquad [2.2]$$

This relative movement can be separated into parts that are normal and tangential to the particles' surfaces at c:

$$d\mathbf{u}^{c,\text{rigid,n}} = du^{c,\text{rigid,n}}\mathbf{n}^c \quad \text{and} \quad d\mathbf{u}^{c,\text{rigid,t}} = d\mathbf{u}^{c,\text{rigid}} - d\mathbf{u}^{c,\text{rigid,n}} \qquad [2.3]$$

where $\mathbf{n}^c$ is the unit contact normal directed outward from p (Figure 1.4); the scalar $du^{c,\text{rigid,n}} = d\mathbf{u}^{c,\text{rigid}} \cdot \mathbf{n}^c$ is the projection of the full movement $d\mathbf{u}^{c,\text{rigid}}$ onto the normal $\mathbf{n}^c$; $d\mathbf{u}^{c,\text{rigid,t}}$ is the projection of $d\mathbf{u}^{c,\text{rigid}}$ onto the tangent plane. Note that a positive value of the scalar increment $du^{c,\text{rigid,n}}$ tends to separate the particles. If the particles are truly rigid and non-penetrating, a non-holonomic constraint applies to the two particles' motions in the form $d\mathbf{u}^{c,\text{rigid}} \cdot \mathbf{n}^c = 0$, which reduces the freedom of motion otherwise available to the particles. Such constraints are the basis of rigid-body methods, such as the Contact Dynamics (CD) simulations described in section 3.4. On the other hand, with Discrete Element Method (DEM) simulations or stiffness matrix methods (sections 3.2 and 3.3), the particles are assumed deformable in the vicinities of their contacts but are otherwise rigid, and the movement $d\mathbf{u}^{c,\text{rigid}}$ is a computational surrogate for the deformation and slip that occur within and between the two particles in the vicinity of their contact.

If the particles or their contacts are not rigid, but can deform and possibly slip, the relative motion in equation [2.2] can be expressed as a sum of deformation and slip contributions:

$$d\mathbf{u}^{c,\text{rigid}} = d\mathbf{u}^{c,\text{def}} + d\mathbf{u}^{c,\text{slip}} \qquad [2.4]$$

where deformation $d\mathbf{u}^{c,\text{def}}$ can be expressed as a sum of normal and tangential components in a manner similar to equation [2.3]. Deformation $du^{c,\text{def}}$ can include both elastic and inelastic contributions, such as micro-slip; whereas, the slip movement $d\mathbf{u}^{c,\text{slip}}$ is entirely tangential and dissipative (frictional). The contact force between the particles will depend on the direction and magnitude of $d\mathbf{u}^{c,\text{rigid}}$, and possibly on the rate $d\mathbf{u}^{c,\text{rigid}}/dt$ and on the history of contact movements (sections 2.2–2.3). The movement $d\mathbf{u}^{c,\text{rigid}}$ satisfies the important property of objectivity: two independent observers moving and rotating independently at uniform rates would measure different translations and rotations of the two particles, but both would report the same relative movement $d\mathbf{u}^{c,\text{rigid}}$. Because of this property, the relative movement can be properly used to compute the resulting contact force.

Equation [2.2] accounts for only three of the twelve degrees of freedom and only three of the six objective degrees of freedom. The other three objective degrees of freedom are associated with rolling-type motions between the particles. Although equation [2.2] unambiguously defines the relative translation, contact rolling can be construed in different ways. A general view of rolling admits alternative forms, but if the force and moment between the particles is to depend on rolling, the rolling definition must be an objective motion that is distinct from the contact translation. In other words, if $d\mathbf{u}^{c,\mathrm{rigid}}$ is zero and the presumed rolling motion is also zero, then the contact force must remain constant, even though the contact force might twirl or rotate in accord with a concerted rigid twirling of the two particles. Such concerted rotations are described near the end of this section (e.g. equations [2.18]–[2.20]).

A reasonable definition of rolling is based on the paths traced by the contact as it moves across the surfaces of the two moving particles. For example, when two inter-meshed gears rotate, two contact points – one on each gear – move from tooth to tooth around the two gears, even as material points on the gears (i.e. their teeth) move in the opposite direction. The rate at which the contact point advances across each of the two surfaces depends on the movements of the two particles and on the curvatures of the surfaces at the contact [MON 88, KUH 04c]. In one definition of rolling, the rolling increment $d\mathbf{u}^{c,\mathrm{roll}}$ is simply the *average* of the increments of the two traced paths: the average of the two increments $\mathbf{t}^{c,p}ds^p$ and $\mathbf{t}^{c,q}ds^q$ across the surfaces of p and q, where the $\mathbf{t}$ vectors are the unit directions of these traced paths within the contact's tangent plane and the ds are the increments' magnitudes. This type of rolling is defined as:

$$d\mathbf{u}^{c,\mathrm{roll}} = \tfrac{1}{2}(\mathbf{t}^{c,p}ds^p + \mathbf{t}^{c,q}ds^q) \qquad [2.5]$$

and the relative tangential movement $d\mathbf{u}^{c,\mathrm{rigid,t}}$ in equation [2.3$_2$] is simply the difference between these two movements:

$$d\mathbf{u}^{c,\mathrm{rigid,t}} = \mathbf{t}^{c,q}ds^q - \mathbf{t}^{c,p}ds^p \qquad [2.6]$$

As the contact point moves across p and q, the direction of normal $\mathbf{n}^c$ will change, driven by rotations of the two particles and by the curvatures of their surfaces. The relationship between changes in the normals $d\mathbf{n}^{c,p}$ and $d\mathbf{n}^{c,q}$ and the displacements $\mathbf{t}^{c,p} ds^p$ and $\mathbf{t}^{c,q} ds^q$ is a problem of differential geometry and is given by:

$$\mathbf{t}^{c,p} ds^p = -(\mathbf{K}^p)^{-1} \cdot (d\mathbf{n}^{c,p} - d\theta^p \times \mathbf{n}^{c,p}) \qquad [2.7]$$

$$\mathbf{t}^{c,q} ds^q = -(\mathbf{K}^q)^{-1} \cdot (d\mathbf{n}^{c,q} - d\theta^q \times \mathbf{n}^{c,q}) \qquad [2.8]$$

where $\mathbf{K}^p$ and $\mathbf{K}^q$ are the surfaces' curvature tensors at the contact point c, and the normals, $\mathbf{n}^{c,p}$ and $\mathbf{n}^{c,q}$, are directed outward from p and q. A smooth curved surface p has two tangential and orthogonal principal curvature directions at a given point: the unit vectors $\mathbf{a}^{1,p}$ and $\mathbf{a}^{2,p}$. For a circular cylinder, one tangent direction, $\mathbf{a}^{1,p}$, is aligned with the cylinder's axis; the other principal direction is in the tangent direction orthogonal to the axis. The principal directions of a sphere are any two orthogonal unit tangent vectors. The scalar curvatures in the two directions, $\kappa^{1,p}$ and $\kappa^{2,p}$, are the inverse radii of curvature in the principal directions with convex surfaces possessing negative values. The principal curvatures of a cylinder of radius ρ are $\kappa^{1,p} = 0$ and $\kappa^{2,p} = -1/\rho$ with zero curvature in the principal direction that is aligned with the cylinder's axis; whereas, the curvatures of a sphere of radius ρ are $\kappa^{1,p} = \kappa^{2,p} = -1/\rho$. When augmented with the outward unit normal direction $\mathbf{n}^p = \mathbf{a}^{3,p}$, the triad of directions $\{\mathbf{a}^{1,p}, \mathbf{a}^{2,p}, \mathbf{a}^{3,p}\}$ constitute the Darboux frame with a pseudo-curvature $\kappa^{3,p} = 0$ assigned to the normal direction. The curvature tensor $\mathbf{K}^p$ is formed from the scalar curvatures and the dyad products of Darboux vectors:

$$K_{ij}^p = \kappa^{1,p} a_i^{1,p} a_j^{1,p} + \kappa^{2,p} a_i^{2,p} a_j^{2,p} + \kappa^{3,p} a_i^{3,p} a_j^{3,p} \qquad [2.9]$$

The normals $\mathbf{n}$ of the two particles p and q and their increments $d\mathbf{n}$ must remain in opposite directions at their contact point c, such that $\mathbf{n}^{c,p} = -\mathbf{n}^{c,q} = \mathbf{n}^c$ and $d\mathbf{n}^{c,p} = -d\mathbf{n}^{c,q}$. Combining these consistency conditions with equations [2.5]–[2.8] leads to an expression for the incremental movement of the contact point (i.e. the increment of the traced path) across the surface of particle p:

$$\mathbf{t}^{c,p} ds^p = -(\mathbf{K}^p + \mathbf{K}^q)^{-1} \cdot \left[(d\theta^q - d\theta^p) \times \mathbf{n}^c - \mathbf{K}^q \cdot d\mathbf{u}^{c,\text{rigid,t}} \right] \qquad [2.10]$$

and a similar expression can be derived for the increment $\mathbf{t}^{c,q} ds^q$ across q:

$$\mathbf{t}^{c,q} ds^q = -(\mathbf{K}^p + \mathbf{K}^q)^{-1} \cdot \left[(d\theta^q - d\theta^p) \times \mathbf{n}^c + \mathbf{K}^p \cdot d\mathbf{u}^{c,\text{rigid,t}} \right] \qquad [2.11]$$

where $d\mathbf{u}^{c,\text{rigid},t}$ is defined in equation [2.6]. As the contact point moves across the surface of p, the direction of the contact normal $\mathbf{n}^c$ also changes, and the incremental change is found by substituting equation [2.10] into [2.7]:

$$d\mathbf{n}^c = d\boldsymbol{\theta}^p \times \mathbf{n}^c + \mathbf{K}^p \cdot (\mathbf{K}^p + \mathbf{K}^q)^{-1} \cdot \left[(d\boldsymbol{\theta}^q - d\boldsymbol{\theta}^p) \times \mathbf{n}^c - \mathbf{K}^q \cdot d\mathbf{u}^{c,\text{rigid},t} \right] \quad [2.12]$$

The increment of rolling, defined by equation [2.5], is:

$$d\mathbf{u}^{c,\text{roll}} = -(\mathbf{K}^p + \mathbf{K}^q)^{-1} \cdot \left[(d\boldsymbol{\theta}^q - d\boldsymbol{\theta}^p) \times \mathbf{n}^c + \tfrac{1}{2}(\mathbf{K}^p - \mathbf{K}^q) \cdot d\mathbf{u}^{c,\text{rigid},t} \right] \quad [2.13]$$

Although this rolling vector is three-dimensional, it lies in the tangent plane of the contact and, thus, is confined to a two-dimensional vector space (i.e. the tangent plane of the two Darboux vectors $\mathbf{a}^1$ and $\mathbf{a}^2$). Equation [2.13] is an objective measure of rolling, since it relies upon the relative rotation of the two particles, $d\boldsymbol{\theta}^q - d\boldsymbol{\theta}^p$, and upon the objective tangential contact deformation $d\mathbf{u}^{c,\text{rigid},t}$. Even though $d\mathbf{u}^{c,\text{rigid}}$ appears in the rolling expression, rolling is distinct from contact deformation: the vectors $d\mathbf{u}^{c,\text{roll}}$ and $d\mathbf{u}^{c,\text{rigid},t}$ are linearly independent, as seen by comparing equations [2.5] and [2.6].

The four equations [2.10]–[2.13] are useful in developing particle simulations and in interpreting the results of physical experiments. Equations [2.10] and [2.11] allow us to track the path of the contact across the surface of each particle; equation [2.12] is useful in assuring that the tangential contact force remains in the tangent plane, even as the particles are rotating and as the contact traverses the surfaces of the particles; equation [2.13] provides an objective measure of rolling that is independent of the contact deformation and can be used in computing the contact force and moment.

The curvature tensor $\mathbf{K}$ that appears in these equations gives the change in the surface normal vector that accompanies an incremental shifting of the surface contact point (equations [2.7] and [2.8]). The 3×3 matrix of Cartesian components $[\mathbf{K}]$ is singular and of rank 2, and the inverses $[\mathbf{K}^p]^{-1}$, $[\mathbf{K}^q]^{-1}$ and $[\mathbf{K}^p + \mathbf{K}^q]^{-1}$ are taken as the corresponding pseudo-inverses $[\,\bullet\,]^+$ defined in equation [1.114]. As such, the rolling vector $d\mathbf{u}^{c,\text{roll}}$ is confined to the two-dimensional space of the contact's tangent plane (see [KUH 04c] for details). The principal curvatures of a convex surface are negative; whereas, the curvatures are positive for a concave surface. The equations are greatly simplified when the surfaces of the two particles are spherical at their point of contact. With radii of curvature ρ^p and ρ^q for two spherical surfaces, the expressions are:

$$\mathbf{t}^{c,p}ds^p = \frac{\rho^p\rho^q}{\rho^p + \rho^q}\left[(d\theta^q - d\theta^p) \times \mathbf{n}^c + \frac{1}{\rho^q}d\mathbf{u}^{c,\text{rigid,t}}\right] \qquad [2.14]$$

$$d\mathbf{n}^c = d\theta^p \times \mathbf{n}^c + \frac{\rho^q}{\rho^p + \rho^q}\left[(d\theta^q - d\theta^p) \times \mathbf{n}^c + \frac{1}{\rho^q}d\mathbf{u}^{c,\text{rigid,t}}\right] \qquad [2.15]$$

$$d\mathbf{u}^{c,\text{roll}} = \frac{\rho^p\rho^q}{\rho^p + \rho^q}\left[(d\theta^q - d\theta^p) \times \mathbf{n}^c + \frac{1}{2}\frac{\rho^p - \rho^q}{\rho^p\rho^q}d\mathbf{u}^{c,\text{rigid,t}}\right] \qquad [2.16]$$

These equations apply to disks, spheres and to cluster-type particles composed of multiple disks and spheres (for example, Figure 3.1f). The last equation is similar to a definition of rolling proposed by Iwashita and Oda [IWA 98] for disks, although they expressed this rolling as an angular increment $d\theta^{c,\text{roll}}$ by dividing the rolling vector $d\mathbf{u}^{c,\text{roll}}$ by a representative radius, taken as the average of ρ^p and ρ^q. Ai et al. [AI 11] present a survey of rolling resistance models that incorporate similar definitions of rolling.

Equations [2.2] and [2.13] account for five of the twelve degrees of freedom of the two particles and five of the six objective motions: three rigid movements $d\mathbf{u}^{c,\text{rigid}}$ and the two tangential rolling components of $d\mathbf{u}^{c,\text{roll}}$. The final, sixth objective motion is the relative "twist" of the particles about the contact normal $\mathbf{n}^c$:

$$d\theta^{c,\text{twist}} = [(d\theta^q - d\theta^p) \cdot \mathbf{n}^c]\mathbf{n}^c \quad \text{or} \quad d\theta_i^{c,\text{twist}} = \left(d\theta_j^q - d\theta_j^p\right)n_j^c n_i^c \qquad [2.17]$$

This twisting motion can produce a small torsional moment between the two particles about the axis $\mathbf{n}^c$ (e.g. [JOH 85]).

Now that all six objective motions are accounted, we should note that other definitions of rolling, other than equation [2.13], can serve in developing particle simulations or in analyzing experimental data. When considered as rolling vectors, these alternative definitions must lie in the vector space of objective motions spanned by equations [2.13] and [2.17]. A simple form of objective rolling is the relative rotation of the two particles, which encompasses both rolling and twisting:

$$d\theta^{c,\text{rel}} = d\theta^q - d\theta^p \qquad [2.18]$$

This definition of rolling is encountered in some DEM codes for calculating a contact moment between particles [TOR 02, JIA 05]. Although this definition of rolling can be used in computing the contact force, we must be aware that when the contact vectors do not coincide, $\mathbf{r}^{c,p} \neq -\mathbf{r}^{c,q}$, the rolling motion is not linearly independent of the relative movement $d\mathbf{u}^{c,\text{rigid},t}$ of equation [2.2]. Bagi and Kuhn [BAG 04] present a third definition of rolling which, as with equation [2.13], is based on a direct averaging of the incremental movements of the contact points $\mathbf{t}^{c,p}ds^p$ and $\mathbf{t}^{c,q}ds^q$ (as in equation [2.5]) [KUH 04d]. For disks and spheres, this rolling definition is the same as equation [2.16].

As a final matter of contact kinematics, we may want to quantify the extent to which two particles twirl or rotate in accord, producing a concerted rigid rotation of the pair. For example, if an entire assembly and its boundary forces are merely rotated as a rigid mass, no deformation occurs at the contacts, but all contact forces must also rotate in concert with the assembly. These particle motions are certainly not objective, since an observer co-rotating with the entire assembly would observe no rotations of the particles or their contact forces. Apart from the rigid rotation of an entire assembly, pairs of particles within a *deforming* assembly can also move, in part, as rigid pairs, and we may wish to distinguish these rigid-like motions from the objective motions that produce contact deformation, rolling, and twisting and the resulting contact forces and moments. Because they are not objective motions, any rigid rotation of a pair of particles must be ignored when computing their contact force, although the contact force must rotate in accord with the pair's rotation. The objective motions, which form a six-dimensional subspace of the twelve degrees of freedom, are distinct from the remaining six-dimensional space of rigid-like motions. Kuhn and Bagi [KUH 05b] derived expressions for these rigid motions, such that the rigid motions span a sub-space that is orthogonal to the space of objective motions. The rigid rotation of a particle pair is as follows:

$$d\theta^{\text{rigid-rot}} = \frac{1}{H}\left[\mathbf{l}^c \times (d\mathbf{u}^q - d\mathbf{u}^p) + 2(d\theta^p + d\theta^q) + \frac{1}{2}\mathbf{l}^c \cdot (d\theta^p + d\theta^q)\,\mathbf{l}^c \right] \quad [2.19]$$

where $H = |\mathbf{l}^c|^2 + 4$, and $\mathbf{l}^c$ is the pair's branch vector (Figure 1.4).

As an alternative to this derivation of the concerted rigid motion of two particles, we can use a combination of the rigid translation (shifting), the rigid tilting and the rigid twirling of two particles. The rigid translation is simply the average of the two particles' translations $\frac{1}{2}(d\mathbf{u}^p + d\mathbf{u}^q)$, and equation [2.12] gives the tilting (rotation) of the contact normal vector $d\mathbf{n}^c$. Besides shifting and tilting, the contact can also undergo a rigid, non-objective twirling rotation about the contact normal, which is simply an average of the two particles' rotations:

$$d\theta^{c,\text{twirl}} = \left[\frac{1}{2}(d\theta^p + d\theta^q) \cdot \mathbf{n}^c \right] \mathbf{n}^c \qquad [2.20]$$

(compare with the objective twisting of equation [2.17]). In numerical simulations, it is essential that the tangential contact force is adjusted to account for this incremental twirling. For example, suppose that two balls are initially engaged with a shearing force between them. If the two particles twirl as a rigid pair about their normal $\mathbf{n}^c$ during the time increment dt with $d\theta^p = d\theta^q = \mathbf{n}^c d\theta$, the normal direction $\mathbf{n}^c$ is unchanged, but the tangential contact force will rotate with the particles in the plane of their contact. (Alternatively, an apparent rotation of tangential force would be seen in a stationary pair of particles when viewed by an observer who is twirling about the direction $\mathbf{n}^c$.) In a similar manner, the normal and tangential forces must also tilt by the amount $d\mathbf{n}$ given in equation [2.12]. This tilting effect is typically performed by projecting the tangential force onto the newly updated tangential plane (for example, [LIN 97]). These rigid rotations are sometimes overlooked.

In advanced contact models, such as those that consider elasto-plastic contact behavior, we must follow the distribution of traction within the contact area – a patch of finite size – from one time step to the next (as in section 2.3.1). This requires that the global motions of the two particles are rotated into a local contact frame and that the resulting contact tractions are then rotated back into the global frame. Such contact models require attention to the cumulative tilt and twirl of the contact [KUH 11].

2.2. Contact mechanics: normal force

Contact mechanics plays a central role in granular geomechanics, as bulk stiffness and strength depend largely on the many forces and movements at individual contacts. As such, behavior observed in laboratory and field tests is merely the bulk expression of the complex interactions of particles at their contacts. Although much effort has been devoted to using numerical simulations to explore bulk behavior, the quality and authenticity of a simulation depends on whether the particle shapes are faithfully represented and on whether the complex and intricate behaviors of the many contacts are credibly characterized by the simulation's idealization of the contact behavior. The small-strain bulk (volumetric) stiffness of a granular material is particularly sensitive to the force–displacement relationship in directions normal to the contacts: the normal (compressive) force f^n and the normal displacement $du^{c,\text{rigid},n}$ (see equation [2.3$_1$]). In the following, we separate a contact force, $\mathbf{f}^c$ or $\mathbf{f}^{pq}$, into normal and tangential components:

$$\mathbf{f}^c = \mathbf{f}^{c,\text{n}} + \mathbf{f}^{c,\text{t}} = -f^{c,\text{n}}\mathbf{n}^c + \mathbf{f}^{c,\text{t}} \qquad [2.21]$$

where the compressive normal force $f^{c,\text{n}}$ and the tangential force $\mathbf{f}^{c,\text{t}}$ are:

$$f^{c,\text{n}} = -\mathbf{f}^c \cdot \mathbf{n}^c \quad \text{and} \quad \mathbf{f}^{c,\text{t}} = \mathbf{f}^c + f^{c,\text{n}}\mathbf{n}^c \qquad [2.22]$$

The relationship between normal displacement and normal force depends on the material properties of the two particles and on the contours and surface textures of the particles at their contact. Two general approaches have been applied to develop displacement–force relationships: methods that ignore surface roughness and attend to the general contours of the contacting surfaces, and methods that focus exclusively on the roughness of surfaces along nearly flat contours. Multiple solutions, summarized below, have been proposed with each approach, and these can be distinguished by whether the solution assumes entirely elastic behavior or admits possible inelastic effects and whether the solution assumes non-adhesive or adhesive contact. This section is focused on the non-adhesive (unbonded) contact.

The simplest form of the force–displacement relation is one in which the normal contact force between two particles is proportional to their indentation (or overlap). This approach has been the basis of most numerical simulations (e.g. sections 2.3.2, 3.2 and 3.3). As seen below, this approach is consistent with elasticity theory for disks but not for spheres, and more advanced contact models are justified for simulations with three-dimensional particles.

2.2.1. *Contact between smooth spheres*

In 1882, Heinrich Hertz solved the problem of frictionless contact between elastic bodies possessing ellipsoidal contours [HER 82], and his approach has been extended to the contact of cylinders, to contact between a conical indenter and a flat surface, to contact between a flat cylindrical indenter and a flat surface, and to other contact topographies [JOH 85, JÄG 05]. The solutions assume that the two bodies are deformed so that their original non-conforming surfaces are reshaped to conform across a small common contact patch. For contact between two spherical contours of equal radius, touching occurs within a plane circular contact area (patch) of radius a, which depends on the normal force between the two spheres and on their elastic properties. In arriving at the solution, the two bodies are usually treated as infinite half-spaces of isotropic and linear elastic material with indentations that are the counterpart of the reshaped surface contours (Figures 2.2(a) and 2.2(b)). When applied to the contact of two particles, it is not necessary that particles are actually spheres, cylinders or some other shapes; the method applies to the *local contours* of the particles in the vicinity of their contact and assumes that the two surfaces are smooth and that the indentation depth ζ and contact radius a are both much smaller than the two particles (Figure 2.2(b)). For contact between two spherical contours, the contact radius a is:

$$a = \sqrt{R\zeta} \qquad [2.23]$$

where indentation ζ is half of the compressive cumulative approach (overlap) of the particles, defined in incremental form in equation [2.3$_1$], such that:

$$\zeta = -u^{c,\text{rigid},n}/2 \qquad [2.24]$$

and R is the contour radius. For spherical surfaces of different radii, R_a and R_b, an averaged radius of curvature applies:

$$R = 2R_aR_b/(R_a + R_b) \qquad [2.25]$$

An important assumption of Hertz theory is that the width of the contact zone is small in comparison with the size of the particles but is large in comparison with their indentation depth (i.e. $\zeta \ll a \ll R$). This assumption leads to deformations within the particles that are localized in the immediate vicinity of the contact and do not extend across the particle bodies.

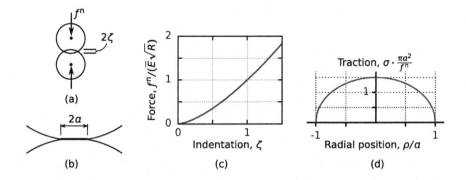

Figure 2.2. *The Hertz solution for contact between two spherical contours: (a) the normal contact force f^n pressing together two particles by distance (overlap) 2ζ; (b) the conforming geometry of the indented surfaces, in which the relationship among ζ, a and R is given in equation [2.23]; (c) the relationship of force and indentation given in equation [2.26]; (d) the profile of the normal traction stress across width $2a$ of the contact area, as in equation [2.28]*

The compressive normal contact force and contact stiffness between two elastic spheres are:

$$f^n = \overline{E}\sqrt{\zeta^3 R} = \overline{E}a^3/R \qquad\qquad [2.26]$$

$$\frac{df^n}{d\zeta} = \frac{3}{2}\overline{E}\sqrt{R\zeta} = \frac{3}{2}\overline{E}a = \frac{3}{2}\left(\overline{E}^2 Rf^n\right)^{1/3} \qquad\qquad [2.27]$$

where modulus $\overline{E}$ is a function of the shear modulus G (or the modulus of elasticity E) and Poisson's ratio v of the two bodies, as $\overline{E} = 8G/(3(1-v)) = 4E/(3(1-v^2))$. The elastic properties of select materials are shown in Table 2.1, and more comprehensive lists are given in [MIT 05]. The Hertz normal force is proportional to $\zeta^{3/2}$, and the stiffness is proportional to $\sqrt{\zeta}$, to the contact radius a or to $(f^n)^{1/3}$. Note that the force f^n and stiffness $df^n/d\zeta$ are continuous and smooth functions of the indentation ζ (Figure 2.2(c)), which are contrary to the usual assumptions in non-smooth models of contact interaction (see section 3.4). The normal contact traction σ within the circular contact area is a function of the radial distance ρ from the center of the area and is given by:

$$\sigma = \frac{3f^n}{2\pi a^3}\left(a^2 - \rho^2\right)^{1/2} \qquad\qquad [2.28]$$

Material	E, GPa	ν	μ	Reference
Glass	71	0.17	0.23	[FUC 14]
Glass			0.4–0.9	[HE 14]
Quartz (wet)	76	0.31	0.4–0.5	[SAN 01, MIT 05]
Gneiss	80	0.27	0.28	[COL 10]
Steel	200	0.27–0.30	0.74	[GRI 97]

Table 2.1. *Elastic and frictional properties of materials*

The normal traction is largest at the center and is zero at the periphery of the contact (Figure 2.2(d)).

A similar method can be applied to contact between two linear-elastic disks (or cylinders) with parallel axes. The half-width a and the normal force f^n are:

$$a = \sqrt{R\zeta} \quad \text{and} \quad f^n = \frac{3\pi}{16}\overline{E}\ell\zeta = \frac{3\pi}{16}\overline{E}\ell a^2/R \qquad [2.29]$$

where ℓ is the disk thickness [JOH 85]. Note that the normal force is a linear function of the indentation, and the contact stiffness $df^n/d\zeta$ is a scalar constant.

2.2.2. *Experiments and pressure dependence of bulk stiffness*

Although Hertz theory is widely used in numerical simulations of granular materials, we must question whether it faithfully represents true behavior as evidenced by experimental results. Tests with indenters possessing relatively smooth spherical surfaces agree reasonably well with the Hertz force in equation [2.26]. For example, Fuchs *et al.* [FUC 14] conducted micro-indenter experiments in which glass spheres of 17 μm diameter were pressed against a flat substrate of silicon with a peak-to-valley asperity roughness of about 1 nm. The fit with Hertz theory was quite close, but only after the diameter was assumed to be 18 μm or after the modulus of the sphere was increased by a factor of 1.05. Indentation experiments with smooth, spherically rounded brass and nylon pins with radii of curvature of about 2 mm showed close conformity with the 3/2 power in equation [2.26] at indentations as small as 0.2 μm [COL 07]. In contact experiments between smooth gneiss spheres with diameter 14.72 mm and surface roughness 173–380 nm, Cole and Peters [COL 10] reported that a 3/2 power holds in the force–deformation relation for deformations as small as 0.1 μm and forces as small as 0.05 N, but they found that the apparent modulus E is less than that measured in separate experiments on bulk specimens. Single particle indentation experiments on irregular gneiss and quartz particles, sand particles and roughened glass spheres also show that contact behavior is softer than the Hertz stiffness at small loads and that a power of less than the 3/2 in equation [2.26] often applies at these loads. Cole and Peters [COL 08] measured the force–deformation behavior of grains of gneiss (both ball milled and natural surfaces), of various lunar simulant materials and of sand. The radius of curvature of the contacts was in the range 0.1–1.0 mm, and they found that a linear (β = 1) relation applied at deformations less than approximately 1 μm. Cavaretta *et al.* [CAV 10] found that borosilicate ballotini spheres with a relatively small roughness (diameter of 1.24 mm and RMS roughness of 0.080 μm) was softer than the Hertz stiffness at normal force less than 4 N but had a stiffness and power relationship similar to Hertz theory at larger loads. Results of similar tests on etched (roughened) ballotini and on sand grains were softer than the Hertz relation at forces less than appoximately 200 N and 25 N, respectively. These results demonstrate that the stiffness at small force decreases with increasing roughness and that a transition to Hertz behavior occurs at larger forces, with the transitional force increasing with increasing roughness.

In regard to the bulk behavior of geomaterials, these experimental results of force and indentation for single- or two-particle tests can be evaluated by considering the particle roughnesses and the average contact forces among the particles within a bulk material. In the experiments cited above, the incremental contact indentations $d\zeta$ were significantly larger than the surface asperities. If we ignore the correlations between contact forces $\mathbf{f}^c$ and branch lengths $\mathbf{l}^c$ in equation [1.53], the average normal contact force f^n within an assembly of equal-size spheres is roughly proportional to the mean stress and the squared particle size:

$$\langle f^n \rangle_{\text{equal spheres}} \approx \frac{\pi(1+e)}{2} \frac{N}{M} pD^2 \qquad [2.30]$$

where p is the (compressive) mean stress, e is the void ratio, D is the particle diameter and N and M are the numbers of particles and contacts within the assembly (the average coordination number is $2M/N$, as in equation [1.15]). For equal-size spheres, the average normal force in equation [2.30] will be in the range of 2–4 times pD^2, and the average contact force between sand-size particles (1 mm size) will be about 0.1 N when the mean stress is about one atmosphere (100 kPa).

By assuming that the bulk strain increment $d\varepsilon$ is roughly proportional to the contact indentations ζ between particles, such that $d\varepsilon \propto d\zeta/D$, evidence of the force–indentation relation can be inferred from the bulk stiffness moduli of granular materials at small strains and from the speeds of compression and shear waves traveling through a bulk material. Wave speed is proportional to the square root of the quotient of the incremental bulk stiffness modulus, say E or G, and the bulk density. Incremental moduli and wave speeds are strongly pressure dependent in granular materials, and this pressure dependence reveals information about the contact force–indentation relation. For a given assembly of particles, the mean stress p (pressure) is roughly proportional to the average of the normal contact forces between particles (see equation [2.30] and the term $\overline{\sigma}^{n,\mathrm{iso}}$ in equations [4.20]–[4.21]), and the stiffness modulus of the bulk material is roughly proportional to the average contact stiffness $df^n/d\zeta$ (see [GOD 90, CHA 91]). By combining these relationships between contact force and mean stress and between contact indentation and strain, the following approximate proportionality for the incremental bulk (volumetric) modulus K^{approx} is as follows:

$$K^{\mathrm{approx}} = \frac{dp}{d\varepsilon_{ii}} \propto \frac{1}{D}\frac{df^n}{d\zeta} \qquad [2.31]$$

Although this relation can be refined by considering the distribution of particle sizes, the particle shapes, the arrangement fabric, the effects of the tangential forces, and the current stress and density conditions, the Hertz result of equation [2.27], when considered in relation to equations [2.30] and [2.31], suggests that the material modulus of the bulk assembly is proportional to p^β with $\beta = 1/3$ and that wave speed is proportional to $p^{\beta/2}$, where p is the mean stress.

In regard to the experiments of Cole and Peters [COL 10] described above, an exponent $\beta = 1/3$ would apply at bulk stresses as low as $(0.05\ N)/(0.0147\ m)^2/4 = 58$ Pa for assemblies of smooth gravel-size spheres of gneiss, which is a rather small stress for most geotechnical applications. To directly test the bulk stiffness of an ideal granular material, Duffy and Mindlin [DUF 57] assembled a bar of unbonded steel balls with a face-centered cubic arrangement and conducted longitudinal resonance tests at different mean stresses. At all stresses, the wave speed was less than that predicted by the Hertz theory. Wave speed increased with increasing mean stress, and at large stresses the value of the exponent $\beta/2$ was about $1/6$, but the exponent was greater than $1/6$ at smaller stresses and was greater for spheres with low dimensional tolerance than for those with higher tolerance (lower variation in diameters). Wave velocity experiments by Otsubo *et al.* [OTS 15] on medium dense specimens of ballotini that had two different surface roughnesses gave an exponent β that decreased with increasing pressure and was larger for specimens with rougher particles. Resonant column experiments on sands generally yield an exponent $\beta/2$ between 0.2 and 0.3 ($\beta = 0.4$–0.6), depending on particle shape, particle size gradation, particle roughness and pre-loading conditions [ARU 92, SAN 98, CHO 06]. Santamarina and Cascante [SAN 98] conducted experiments on irregular assemblies of mono-disperse spheres with different surface conditions and, in agreement with the other studies described above, found that the exponent increased with surface roughness. Based on these and other experiments and field studies, a value $\beta = 0.5$ is widely used in geotechnical practice for analyzing the propagation of seismic waves and for predicting earthquake ground motions and building settlements, for the range of confining stress that commonly applies in these situations.

DEM simulations of sphere assemblies with Hertz contacts also give bulk stiffness moduli that depend on mean stress, but with an exponent β greater than $1/3$ [AGN 07b]. The author's simulations of sphere assemblies yield a value $\beta = 0.42$, and simulations on assemblies of bonded sphere-cluster yield $\beta = 0.39$ [KUH 14c]. These values, which exceed the expected results of $1/3$, are due to non-uniform distributions of contact force, to particle rotations, and to the effects of the tangential contact forces [AGN 07b]. In any event, the exponents β from DEM simulations with Hertz contacts are generally greater than $1/3$ but are less than the experimental value 0.5 that is commonly measured with bulk sand specimens.

2.2.3. *Contact between non-spherical contours*

To correct this deficiency in DEM simulations, we can extend Hertz theory to non-spherical asperities, rather than treating the surface contours of two particles as spheres with a diameter equal to the size of the particles themselves. Jäger [JÄG 99] derived the normal force f^n between two surfaces, each possessing a power-law contour of general form: a surface of revolution with $z = A_\alpha r^\alpha$ (see Figure 2.3(a)). The result is a suite of force–displacement relations with exponents $\beta\ (= 1 + 1/\alpha)$ other than the 3/2 of spherical contours:

$$f^n = C_\alpha \zeta^{1+1/\alpha}, \quad C_\alpha = \frac{4\alpha G}{(1-v)(1+\alpha)} \left(\frac{\Gamma\left(\frac{1+\alpha}{2}\right)}{\sqrt{\pi} A_\alpha \Gamma\left(\frac{2+\alpha}{2}\right)} \right)^{1/\alpha} \qquad [2.32]$$

where α is the shape parameter, ζ is the indentation depth (half of the contact overlap), G and v are the shear modulus and Poisson's ratio of the grains and Γ is the gamma function. The force–displacement relation, therefore, is altered by the contour parameter α. Smaller values of α correspond to sharper profiles, and larger numbers giving blunter shapes (Figure 2.3(b)). For smooth spherical surfaces of radius R, the exponent α is 2 and the contour parameter A_2 is $1/(2R)$, so that equation [2.32] yields the standard Hertz solution of equation [2.26].

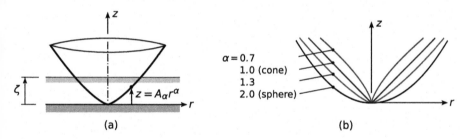

Figure 2.3. *Contours of contact asperities as power-law surfaces of revolution: (a) general power-law form with exponent α and parameter A_α; (b) profiles for a range of α exponents*

Noting that contact asperities can have non-spherical profiles, Goddard [GOD 90] proposed modeling contacting bodies as elastic cones that meet tip-to-tip. With a conical asperity ($\alpha = 1$, Figure 2.3(b)),

$$C_1 = \frac{4G_s}{\pi(1-v_s)} \frac{1}{A_1}, \quad f^n = \frac{4G_s}{\pi(1-v_s)} \frac{1}{A_1} \zeta^2 \qquad [2.33]$$

in which A_1 corresponds to the outer slope of the cone. The corresponding contact stiffness $df^n/d\zeta$ is proportional to ζ and to the square root of the normal force $(f^n)^{1/2}$. By decoupling the asperity shape from the more general contour of a particle's surface, equations [2.32] and [2.33] afford a free parameter A_1 that can be calibrated so that the granular assembly has bulk moduli similar to that of a targeted granular material. When incorporated in a DEM simulation of assemblies of spheres and of sphere clusters, a conical surface with $\alpha = 1$ results in a pressure dependence of $\overline{E}$ with an exponent $\beta = 0.56$, which is close to the accepted value of 0.50 [KUH 14c]. The author regularly uses the factor $\alpha = 1.3$ in DEM simulations, which produces a bulk exponent $\beta = 0.50$.

2.2.4. *Contact between rough contours*

Rather than treating contact as occurring within a single, compact area (patch) between the curved surfaces of two particles, an alternative approach is to treat the surfaces as nominally flat but to also admit the presence of fine-scale surface roughness in the form of numerous irregular asperities. Figure 2.4 shows the surface contour of an apparently mirror-smooth specimen of quartz, as revealed with a diamond stylus profilometer [LAM 69]. This equipment can identify the "peaks" encountered as the profilometer probe moves in a line, although even higher "summits" can exist elsewhere across the full surface. Surface roughness can be quantified with the RMS deviation of the surface from its mean profile (about 50 nm for the surface in Figure 2.4). Although the surface deviation is commonly assumed as Gaussian in its distribution, some evidence suggests that roughness has a fractal character, such that roughness of increasingly finer (but self-similar) detail is revealed with each increase in resolution.

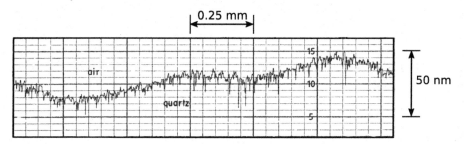

Figure 2.4. *Profile of a smooth quartz surface (note the two scales) [LAM 69]*

When rough particles are pressed together, they first touch at high spots along their surfaces. With increasing load, existing contacts between asperities increase in area and the gaps between other asperities close and touch, contributing to the total contact area. Because the radii of curvature of the asperities are much smaller than the radius of the mean surface profile, the contact pressures between asperities are smaller than that of Hertz contact between smooth spherical surfaces, but the full area of contact between asperities is greater than that of a smooth sphere profile. However, because of the inter-asperity gaps, the combined contact area among the asperities is less than that of the full projected contact area (the nominal area). Greenwood and Williamson [GRE 66] idealized the contact between nominally flat surfaces as occurring at elastic asperity summits with spherical tips. By assuming a uniform tip radius with a Gaussian distribution of heights, they found that the combined contact area was nearly proportional to the applied normal load, a confirmation of the Bowden and Tabor experiments (see page 84) [BOW 73]. This proportionality is due to the average radius of the summit contact areas remaining constant with increased loading: existing contacts are broadened but new contacts are engaged as the surfaces approach each other, increasing the number of contacts of an unchanging average size [JOH 85]. Because of this constant average radius of contact among an increasing number of contacts across a flat surface, the average contact stress σ remains constant as the load increases, and the cumulative stiffness $df^n/d\zeta$ is proportional to f^n, which leads to bulk behavior in which the stiffness modulus is proportional to the mean stress (a β of 1.0, as defined in the previous sections).

These results for the contact of flat surfaces with asperities must be modified for curved, contoured surfaces. By analyzing contact between a rough spherical surface and a smooth sphere, Greenwood and Tripp [GRE 67] derived an equivalent radius of contact a^* for a given normal force and roughness. With small forces, the equivalent radius is much greater than the contact radius a of the smooth profile, but as the force increases, the equivalent radius a^* approaches a. Yimsiri and Soga [YIM 02] applied the Greenwood and Tripp model to sand particles 1 mm in size with an RMS asperity height of 3.5 μm and showed that the contact stiffness exponent β transitioned from a value of 0.71 at normal forces less than 1.5 N to the Hertz value of $1/3$ at forces of above 100 N.

2.3. Contact mechanics: tangential force

Many unusual aspects of the load–deformation behavior of granular materials originate from the frictional character of the particle interactions, which places a limit on the magnitude of the tangential contact force, $|\mathbf{f}^t|$, between two particles:

$$|\mathbf{f}^t| \leq \mu f^n \qquad\qquad [2.34]$$

where μ is the coefficient of friction (values of μ for select materials are shown in Table 2.1, and more comprehensive lists are given in [MIT 05]). This simple rule, called Amontons' first law, was first noted by Leonardo da Vinci, rediscovered by Guillaume Amontons, and further developed by Charles-Augustin de Coulomb. The energy dissipation and inelastic behavior of granular materials are largely attributed to the frictional sliding at contacts. The coefficient μ is empirically derived and depends on the particles' mineralogy, the surface condition (roughness, interstitial fluids, chemical contamination, etc.), sliding velocity, wear of the sliding surfaces, temperature and other factors. When these factors are controlled, equation [2.34] remains valid for a wide range of normal forces, contact areas and particle sizes, provided that any attractive forces between the particles (particularly between micron-size particles) are included in the normal force f^n. The observation that the friction limit is independent of contact area is referred to as Amontons' second law. An independence of μ and velocity, which is contrary to a viscous origin of friction, is called Coulomb's law. Numerous explanations have been offered for the linear nature of the friction limit, but the underlying physics is still a matter of study, and a satisfactory explanation from first principles is still an open question [GAO 04]. By measuring electrical conductivity across contacting bodies, Bowden and Tabor [BOW 73] discovered that the actual combined contact area between asperities is proportional to the contact force, unlike the predictions for a Hertz contact, in which the area is proportional to $(f^n)^{2/3}$ (see equation [2.26]). They proposed that the friction limit is reached through plastic flow within the asperities, and that the linear increase in the asperity area with an increase in normal force is the fundamental basis of the proportionality in equation [2.34].

The linear relation between the friction force and the normal force that had been found between wood blocks by da Vinci [DOW 79] also applies at the scale of nano-size contacts. Gao *et al.* verified this linearity using a friction force microscope (FFM) by sliding silicon tips of radii 11 and 33 nm across gold surfaces, each giving the same friction coefficient μ = 0.49 ± 0.02 [GAO 04]. Fuchs *et al.* [FUC 14] used atomic force microscopy (AFM) to measure friction between borosilicate glass spheres of 20 nm radius and flat silicon wafer substrates. The results conformed the linear equation [2.34], although μ depended on the surface roughness of the substrate, which ranged across RMS values of 0.3–2.7 nm. At a much larger scale, Cole *et al.* [COL 10] measured the friction limit of contact between spherical gneiss surfaces with diameters 14.72 mm, and they found that the limit is proportional to normal force for forces in the range 2–20 N. In tests on steel, glass ballotini and natural sand particles, Horn and Deer [HOR 62] and Cavaretta *et al.* [CAV 11] found that μ increases with surface roughness and is much larger for wet than for dry particles.

Equation [2.34] establishes a limitation on the tangential force, but the relationship between force and movement for smaller tangential forces will influence the bulk stiffness and deformation. Unlike the normal force between elastic bodies, which can be expressed as a unique relationship between force and accumulated indentation ζ (for example, equation [2.26]), the relationship between tangential force and movement is an *incremental* and history-dependent relationship. Two common models of tangential force are considered in this section: an incrementally nonlinear model that is the complement of the Hertz normal force, and a much simpler linear-friction model.

2.3.1. *The Cattaneo-Mindlin model*

Although the friction limit described in equation [2.34] originates within the contacting surfaces and in the asperities that touch within the contact area, the load–displacement behavior for tangential loads below this limit is the result of deformations within the particle bodies as well as more localized deformations within the much smaller asperities. Cattaneo [CAT 38] and Mindlin [MIN 49] studied tangential loading between linear-elastic spheres as a counterpart to Hertz theory for normal forces (section 2.2).

According to Hertz theory, the radius a of the circular contact area between two spheres depends entirely on the normal force f^n. The distribution of normal traction σ (i.e. equation [2.28] and Figure 2.2(d)) also depends exclusively on the normal force and is not changed by the tangential force. As with the normal traction, the tangential traction τ in Cattaneo-Mindlin theory is radially symmetric and is a function of radial position ρ relative to the center of the contact area. The tangential traction, however, is more complex than the normal traction, since the ratio $|\tau|/\sigma$ of tangential and normal tractions can nowhere exceed the limit μ. Reaching this limit results in micro-slip, which occurs within those portions of the contact area where $|\tau| = \mu\sigma$, even as the two spheres adhere within the remainder of the contact area, where $|\tau| < \mu\sigma$. With each new increment of normal and tangential force (or of normal and tangential movement), the previous distribution of tangential traction is altered, but because of the frictional limit, the alteration of tangential traction is not in proportion to the force increments. The friction limitation on the tangential traction, therefore, leads to path-dependent inelastic behavior. Mindlin and Deresiewicz [MIN 53] analyzed 11 of the infinite variety of loading paths and concluded that the tractions and the load–displacement stiffness depends not only "on the initial state of loading, but also on the entire past history of loading and the instantaneous relative rates of change of the normal and tangential forces," and on the directions and relative magnitudes of the increments of normal and tangential movement.

As the simplest example of a complex problem, Mindlin and Deresiewicz analyzed a scenario in which two equal-size spheres are first pressed together with normal force f^n, and then the tangential force is monotonically increased from zero to f^t while maintaining constant normal force (the path shown in Figure 2.5(a)). Displacement ξ is the tangential displacement of the center of the contact area relative to the center of one sphere (Figure 2.5(b)). That is, the increment $d\xi$ is half the relative movement $du^{c,\text{rigid,t}}$ in equations [2.2], [2.3$_2$] and [2.6]. If the entire circular contact area was to translate in a rigid manner in the direction of f^t, the tangential traction would be infinite at the periphery of the area. Instead, frictional slip (micro-slip) begins at the outer extremity of the circular contact area, creating a thin peripheral annulus of micro-slip, while no slip occurs within a central circular region (Figures 2.5(c) and 2.10(a)). As the tangential force increases, the annular region of micro-slip increases in area: the outer radius a, established by the normal force, remains unchanged, but the inner radius c of the micro-slip region shrinks inward. Within this central portion, the two spheres remain in non-slip firm contact, such that this circular non-slip area translates as a rigid surface in the directions of $d\xi$ and $\mathbf{f}^t$. The radial distribution of tangential shearing traction τ is plotted in Figure 2.5(c) for this particular loading path, with the plot's origin at the center of the circular contact area of radius a. As radius c of the central non-slipping region shrinks, the tangential stiffness is reduced, and slip across the full region finally occurs when the friction limit of equation [2.34] is reached.

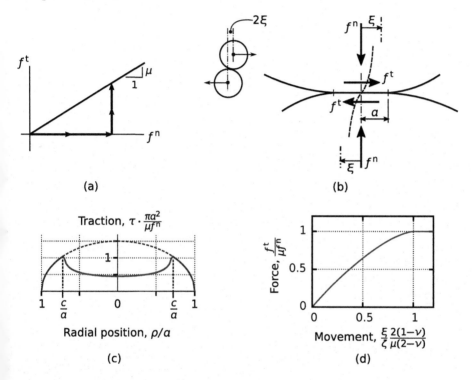

(a)

(b)

Traction, $\tau \cdot \dfrac{\pi a^2}{\mu f^n}$

Radial position, ρ/a

(c)

Force, $\dfrac{f^t}{\mu f^n}$

Movement, $\dfrac{\xi}{\zeta}\dfrac{2(1-\nu)}{\mu(2-\nu)}$

(d)

Figure 2.5. *Results of Cattaneo-Mindlin theory for the tangential loading of two equal-radius elastic spheres: (a) example loading path of normal and tangential forces; (b) tangential movement ξ and force f^t; (c) tangential traction within the circular contact area of radius a for the loading path in (a) with micro-slip within $c/a < \rho/a < 1$; (d) load and displacement behavior for the loading path in (a)*

Figure 2.5(d) shows the tangential force–displacement behavior for this example of monotonically increasing tangential force under constant normal force, and full slip occurs when the tangential movement ξ equals the indentation ζ (half the overlap) times $\mu(2 - \nu)/(2(1 - \nu))$. The tangential stiffness for the simple loading path in Figure 2.5(a) is:

$$\frac{df^t}{d\xi} = \frac{3}{2}\frac{2(1 - \nu)}{2 - \nu}\overline{E}\,a\left(1 - \frac{f^t}{\mu f^n}\right)^{1/3} \qquad [2.35]$$

where the material modulus $\overline{E}$ is defined below equation [2.27]. Comparing the normal and tangential stiffnesses in equations [2.27] and [2.35], the maximum tangential stiffness – at the initial application of tangential force – is smaller than the normal stiffness by the factor $2(1 - \nu)/(2 - \nu)$ and is reduced to zero at the friction limit. In some simulation codes [CUN 88a, LIN 97], the tangential stiffness is not degraded in this manner with further tangential load, but instead, the stiffness is assumed independent of f^t and of the loading history and is assumed equal to this factor multiplied by the normal stiffness in equation [2.27].

Johnson [JOH 55] recorded the tangential movement and force of a steel ball against a flat steel plate under constant normal force, and the results agreed closely with Cattaneo-Mindlin theory (i.e. Figure 2.5(d)). In further tests, he oscillated the tangential movement while increasing its magnitude and found that the resulting hysteresis loops of displacement and force agreed with the theory of Mindlin and Deresiewicz (Figure 2.6). The area inside these loops represents frictional energy dissipation due to micro-slip within the outer, annular contact area, even as the central part of the area remains in non-slip, adhering contact. Other experiments by Goodman and Brown [GOO 62] where steel spheres were pressed against a flat steel surface verified the Mindlin and Deresiewicz result of frictional dissipation in each hysteresis cycle. The results of sphere–sphere and sphere–flat experiments by Cole et al. [COL 10] with smooth specimens of gneiss agreed in a qualitative way with Cattaneo-Mindlin theory: the tangential stiffness increases with increasing normal force, cyclic loading produces hysteresis in the tangential force–displacement response and the tangential stiffness generally decreases as the tangential force increases toward the condition of full slip. During similar tangential loading, Senetakis et al. [SEN 13] also measured a reduced tangential stiffness with specimens of quartz and limestone in the size range 1.18–5.00 mm.

In the solution of the particular Cattaneo-Mindlin problem shown in Figure 2.5(a), micro-slip occurs when the normal force is constant but tangential movement is monotonically advanced. No slip occurs, however, when the normal indentation ζ is increased in unison with a monotonic advance in the tangential displacement ξ, provided that the ratio $|d\xi|/d\zeta$ is less than the friction coefficient μ multiplied by the factor $(2 - v)/(2(1 - v))$. When an elastic movement of this type is followed by a tangential movement with no further change in the normal force, the resulting tangential force is affected by the ratio $d\xi/d\zeta$ and the extent of the preceding elastic step. Unlike the linear-frictional model of section 2.3.2, micro-slip can occur at any stage of loading of a Cattaneo-Mindlin contact. The contact behavior is incrementally nonlinear, and regardless of the current contact force, at least two branches of loading are available: an elastic branch and a micro-slip branch. Moreover, when the friction limit is reached with $|\mathbf{f}^t| = \mu f^n$, three branches are available: elastic, micro-slip and full slip of the contact.

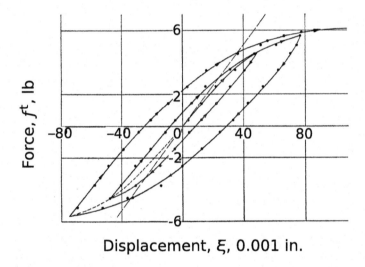

Figure 2.6. *Results of Johnson [JOH 55] for oscillatory tangential displacement of a steel ball across a flat steel surface with constant normal force. Normal force f^n = 13.9 lb and radius R = 0.188 in*

Mindlin and Deresiewicz [MIN 53] had solved the displacement–force relation for several loading paths involving two or three loading steps, and they proposed an approach for extending their approach to more complex paths. With DEM simulations (section 3.2), an unambiguous algorithm is required for computing the tangential force for arbitrary increments in the normal and tangential movements, $d\zeta$ and $d\xi$. With each of their 11 loading sequences, Mindlin and Deresiewicz [MIN 53] only analyzed movements ξ in a single direction, although more general increments $d\mathbf{f}^t$ and $d\xi$ are possible within the tangential plane. An algorithm that fully accounts for the history-dependent response must integrate the full sequence of movements – starting from the moment that two particles first touch and finishing with the most recent movement increment. A comprehensive method must also allow for tangential movements (and sequences of movements) in all tangential directions. Seridi and Dobry [SER 84] approached this difficult problem by analyzing an additional loading possibility in which tangential motion in one direction is abruptly followed by an infinitesimal tangential movement in the perpendicular direction. Placed in the context of an elasto-plastic continuum, they found that the results resembled a loading probe tangent to a yield surface. They also developed a general algorithm for the three-dimensional displacement–force relation that was analogous to incremental elasto-plasticity with kinematic hardening [DOB 91]. Vu-Quoc et al. [VUQ 04] developed an alternative tangential force–displacement algorithm that uses an incremental form of the Mindlin-Deresiewicz stiffness, and they applied the model to four combinations of tangential and normal force increments that are each either increasing or decreasing.

The limitations and computational difficulties of these methods were largely circumvented by Jäger [JÄG 05], who expressed an otherwise complex distribution of shearing tractions as a superposition of simple Cattaneo-Mindlin functions, which provides a compact means of chronicling the essential elements of a complex loading history. The algorithm [KUH 11] computes the contact force for each step of an arbitrary sequence of normal and tangential movements without the need to divide a large movement into smaller, incremental sub-movements, while reproducing the closed-form solutions of Mindlin and Deresiewicz. Figure 2.7 shows an application of the Jäger algorithm to a complex sequence of normal and tangential displacements that increase and decrease in various combinations. The complexity of the solution is seen in the final distribution of tangential traction, in which the various spikes correspond to past events of load reversal and micro-slip (Figure 2.7(b)). In the Jäger approach, the history of these load reversals is captured in an equivalent loading path, which is amended with each loading increment.

With the simpler linear-frictional model of the next section, a distinction is made between contact movements that produce elastic deformation and movement that produces slip. Such distinction is somewhat ambiguous for a Cattaneo-Mindlin contact model, as deformation and slip can occur together, and slip can occur in the form of micro-slip or as full sliding of the particles. Because the deformation movement $d\mathbf{u}^{c,\text{def}}$ can include both elastic and plastic contributions, micro-slip can be considered part of $d\mathbf{u}^{c,\text{def}}$ (equation [2.4]). The elastic (recoverable) part of contact movement can be computed by tracing movements that had produced the Jäger equivalent loading path.

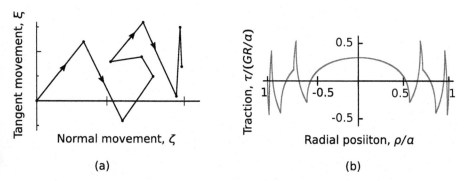

(a)   (b)

Figure 2.7. *Complex sequence of tangential and normal movements: (a) movement sequence; (b) final radial distribution of shear traction across the circular area ($\mu = 0.5$, $v = 0.5$, $\overline{E} = 5.3$ and $R = 1$). Compare with Figure 2.5(c).*

2.3.2. Linear-frictional model

This simple model of tangential contact behavior assumes linear and elastic displacement–force behavior until the friction limit is reached, after which the contact undergoes full slip. The model was reviewed by Michalowski and Mroz [MIC 78] and was the original contact model used with the discrete element method (DEM) [CUN 79]. Although more advanced models have been applied in DEM simulations, the linear-frictional model remains the most frequently used.

Prior to slip, both the normal and tangential behaviors are linear and elastic:

$$\frac{df^{n}}{d\zeta} = k^{n} \quad \text{and} \quad \frac{df^{t}}{d\xi} = k^{t} = \alpha k^{n} \tag{2.36}$$

where k^t and k^n are the tangential and normal stiffnesses and α is their ratio. Overlaps (twice ζ and ξ) are conventionally used in numerical simulations, and the corresponding stiffnesses are half those in equation [2.36]. When the frictional limit is reached (when $|\mathbf{f}^t| = \mu f^n$), the contact stiffness is incrementally nonlinear with two possible branches: an elastic branch that is characterized by the normal and tangential stiffnesses k^n and αk^n, and a sliding branch characterized by the friction coefficient μ. The active branch is determined by the direction of the contact deformation $d\mathbf{u}^{\text{rigid}}$, with its normal and tangential components $2d\zeta = du^{\text{rigid,n}}$ and $2d\xi = d\mathbf{u}^{\text{rigid,t}}$ (see equations [2.2] and [2.3]). Sliding can only occur when two conditions are met:

1) When the current contact force satisfies the yield condition $Q = 0$, as embodied in the friction limit of equation [2.34]:

$$Q = Q(\mathbf{f}) = |\mathbf{f} - (\mathbf{n} \cdot \mathbf{f})\,\mathbf{n}| + \mu \mathbf{f} \cdot \mathbf{n} \tag{2.37}$$

where $\mathbf{f}$ is the three-dimensional contact force vector (see equations [2.21] and [2.34]). The yield surface Q is a cone in the force-space with an axis aligned with the direction of the contact normal $\mathbf{n}$.

2) When the contact deformation $d\mathbf{u}^{\text{rigid}}$, defined in equation [2.2], is directed outward from the yield surface in displacement space, the condition $S > 0$:

$$S = S(\mathbf{f}, d\mathbf{u}^{\text{rigid}}) = \mathbf{g} \cdot d\mathbf{u}^{\text{rigid}} > 0 , \tag{2.38}$$

where the yield surface Q has the normal direction

$$\mathbf{g} = k^n (\alpha \mathbf{h} + \mu \mathbf{n}) \tag{2.39}$$

and the unit sliding direction $\mathbf{h}$ is tangent to the contact plane and aligned with the current contact force $\mathbf{f}$:

$$\mathbf{h} = \frac{\mathbf{f} - (\mathbf{n} \cdot \mathbf{f}) \cdot \mathbf{n}}{|\mathbf{f} - (\mathbf{n} \cdot \mathbf{f}) \cdot \mathbf{n}|} \tag{2.40}$$

With this simple model and its hardening modulus of zero, the contact stiffness can be expressed in matrix form with the incremental stiffness matrix $\mathbf{K}$,

$$d\mathbf{f} = \mathbf{K} \cdot d\mathbf{u}^{\text{rigid}} \tag{2.41}$$

The matrix has two branches, elastic and sliding, given by:

$$\mathbf{K} = \frac{1}{2} \begin{cases} \mathbf{K}^{\text{elastic}} = k^{\text{n}} \left[\alpha \mathbf{I}_3 + (1 - \alpha) \mathbf{n} \otimes \mathbf{n} \right] & \text{if } Q < 0 \text{ or } S \leq 0 \\ \mathbf{K}^{\text{slip}} = \mathbf{K}^{\text{elastic}} - \mathbf{h} \otimes \mathbf{g} & \text{if } Q = 0 \text{ and } S > 0 \end{cases} \qquad [2.42]$$

where $\mathbf{I}_3$ is the 3×3 identity matrix δ_{ij}. The factor of ½ in this stiffness results from $d\mathbf{u}^{\text{rigid}}$ in equation [2.41] being twice the increments $d\zeta$ and $d\xi$ in equation [2.36]. Note that equations [2.41]–[2.42] give both the normal and tangential force components. Because the sliding and yield directions do not coincide ($\mathbf{h} \neq \mathbf{g}$), sliding is non-associative and the stiffness $\mathbf{K}$ in equation [2.42$_2$] is asymmetric and may lead to negative second-order work at the contact. The sliding behavior possesses deviatoric associativity; however, since the sliding direction $\mathbf{h}$ is aligned with the tangential component of the yield surface normal $\mathbf{g}$ [BIG 00]. Another aspect of equation [2.42] is that a linear-frictional contact has an incrementally nonlinear behavior with both an elastic branch and a slip branch, when the yield (slip) condition is attained, $Q = 0$.

2.4. Contact mechanics: rolling resistance

In section 2.1, several definitions were given for the relative rolling motion between two particles (see equation [2.13] and its following paragraphs), and resistance to such rolling may develop at the particle contacts during the bulk deformation of a granular material. Several models of rolling resistance have been implemented in numerical simulation codes, usually by applying a contact moment to complement the contact force. A different form of rolling resistance is encountered in problems of tire–pavement and wheel–rail tribology. These two forms of rolling resistance can be distinguished by considering a pair of smooth cylindrical rollers of equal radius, one being driven by the other (Figure 2.8). Various mechanisms can lead to energy dissipation, causing the input power $w_1\dot{\theta}_1$ to exceed the output power $w_2\dot{\theta}_2$, $\dot{W}_{\text{input}} \geq \dot{W}_{\text{output}}$ (here, w and $\dot{\theta}$ represent externally applied moments and rotational velocities, as shown in Figures 2.1 and 3.4). With the first category of rolling resistance, a contact moment acts between the two rollers so that the input moment exceeds the output moment, $|w_1| > |w_2|$, yet preserving the rotational velocities, $|\dot{\theta}_1| = |\dot{\theta}_2|$. This type of rolling resistance, called a *rolling moment*, was introduced in the DEM simulations of Iwashita and Oda [IWA 98] and is now available in many simulation codes (see Ai *et al.* [AI 11] for a survey) and is further described below. The presence of a contact moment is most obvious in the interaction of two gears that rotate with synchronous velocities, but in which the teeth rub in manner that causes a torque reduction between the two gears – a reduction that can be modeled as a contact moment (section 2.4.1). With the second mechanism, the absence of a contact moment preserves the torque (i.e. $|w_1| = |w_2|$), but the output rotational *velocity* is diminished between the driving roller and the driven roller, $|\dot{\theta}_1| > |\dot{\theta}_2|$. This latter type of resistance, described in section 2.4.2, is termed *rolling friction* or *creep-friction* and results from micro-slip between the two rollers (or two particles) within their small contact region, causing energy dissipation in the absence of a contact moment.

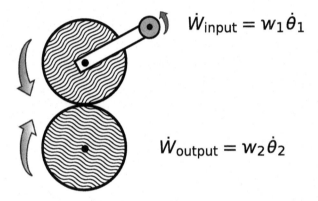

$$\dot{W}_{input} = w_1\dot{\theta}_1$$

$$\dot{W}_{output} = w_2\dot{\theta}_2$$

Figure 2.8. *Input and output power, $\dot{W}_{input}$ and $\dot{W}_{output}$, with*
$w_1 \cdot \dot{\theta}_1 \geq w_2 \cdot \dot{\theta}_2$ *[KUH 14b]*

2.4.1. *Rolling moment*

This first type of rolling resistance is the result of a net contact moment $\mathbf{m}^c$ that develops within the contact area between two particles (Figure 1.7(b)). Rolling moments can give rise to an asymmetry of the bulk stress and can justify the continuum representation of a bulk material as a micro-polar (Cosserat) medium (section 1.3.2 e.g. equations [1.57], [1.78] and [1.84]). The inclusion of contact moments has become common in discrete element (DEM) simulations [AI 11] and more recently in contact dynamics (CD) simulations [HUA 13], although a rolling moment is usually invoked in simulations with circular and spherical particles to produce results that resemble those of irregularly shaped particles. In this manner, rolling moments are used as an expedient surrogate to capture the effects of particle shape, rather than to model any actual moments at the contacts. Some physical processes, however, will give rise to true rolling moments.

For an elastic Hertz contact between two spheres, the normal traction σ within the circular contact area between the particles is radially symmetric and is given by equation [2.28]. Such symmetry obviates a contact couple, and a rolling moment can only exist under conditions that break this symmetry. These conditions include (1) plastic and visco-elastic deformations of the particles near the contact, (2) adhesion or cementation at the contact. As two particles roll across each other, the contact area will move across the particles, freshly engaging new material at the leading edge of the contact while disengaging previously contacting material at the trailing edge. Because the leading and trailing parts of the contact have a different stress history, a contact's normal tractions will differ for its leading and trailing portions if the material undergoes plastic or visco-elastic hysteresis [HAM 63, BRI 98]. Such inelastic deformation can occur within the bodies of the particles or within their surface asperities. As an approximation, the contact moment m will equal the imbalance in normal traction f^n between the leading and trailing parts multiplied by the contact radius a. Taking this imbalance as the product of a factor μ^{roll} and the normal force, the contact moment is:

$$m = \mu^{roll} f^n a \qquad\qquad [2.43]$$

If we also approximate the contact width as that of a Hertz contact (equations [2.23] and [2.26]), then the contact moment is proportional to the fourth power of the width, a^4, and to the contact indentation squared, ζ^2. This moment, although small, might be significant when the force imbalance $\mu^{roll} f^n$ is large when compared with the applied stress or the self-weight of the particles. These conditions can apply to particles of colloid size.

A force imbalance can also occur when particles adhere: either the result of vigorous cementation bonding or of attractive forces between the particles, such as van der Waals and electrostatic attractions. Johnson, Kendall and Roberts (JKR) [JOH 55] derived the contact radius of two elastic spheres attracted through an adhesive (tensile) surface energy γ and repelled with the

Hertz (compressive) traction, but with no external force pressing them together:

$$a_{\mathrm{JKR}}^3 = \frac{12\pi\gamma R^2}{E} \qquad [2.44]$$

The net pull-off force P_c is related to the surface energy, as $P_c = -(3/2)\pi\gamma R$. When two particles roll, an asymmetric traction occurs across the contact area, with fresh adhesion occurring along the leading edge and tensile pealing taking place along the trailing edge (Figure 2.9(b)). Dominik and Tielens [DOM 95] applied JKR theory to determine the moment stiffness between attractive spheres of radius R, and this stiffness depends on the surface energy γ or the pull-off force P_c, and is approximated as:

$$\frac{dm}{d\theta} = 6\pi\gamma R^2 = 4P_c R \qquad [2.45]$$

where $d\theta$ is the relative rotation of the spheres. Note that this stiffness is quite small for particles larger than colloidal size.

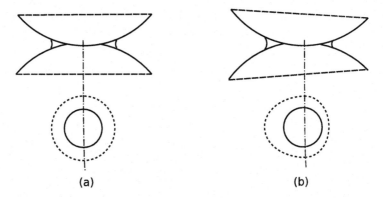

(a)   (b)

Figure 2.9. *Contact moment produced between two adhesive spheres: (a) symmetric contact between two spheres prior to rotation. Solid circle is the Hertz contact area of radius a, and the outer dashed circle represents the zone of adhesive attraction; (b) asymmetric contact after the two particles undergo a relative motion, which produces a counteracting contact moment*

Ding *et al.* [DIN 07] conducted rolling stiffness experiments on polystyrene latex spheres of 22.4 μm diameter on a flat silicon substrate. Prior to reaching the rolling friction limit, the stiffness $dm/d\theta$ ranged from 2 to 7×10^{-11} N-m. Fuchs *et al.* [FUC 14] measured the rolling friction between 10 μm spheres of borosilicate glass on a flat silicon substrate. Contrary to sliding friction, the rolling coefficient μ^{roll} decreased with increasing roughness of the substrate, with $\mu^{\text{roll}} = 7.6 \times 10^{-3}$ for the smoothest surface (0.3 nm RMS roughness). The rolling coefficient, as small as it was, was derived from adhesion between the sphere and substrate, as the pull-off force (3.2 μN) was several orders of magnitude greater than the particle's weight. Such rolling moments will be insignificant; however, unless the spherical particles are of less than 1 millimeter in size and are confined with a mean stress that is less than their self-weight at burial depths of a few millimeters.

2.4.2. *Creep-friction*

As with the Cattaneo-Mindlin friction model described in section 2.3, creep-friction results from micro-slip within the contact area of two touching particles. In Cattaneo-Mindlin sliding of two spheres (Figure 2.10(a)), the contact area is assumed to remain at the same location on the spheres' surfaces, with the two particles firmly adhering within a central non-slip area. Micro-slip occurs in an outer annular region, and it is only when the full frictional limit is reached that the entire contact area begins to traverse the surfaces of the two particles. With creep-friction (Figure 2.10(b)), the contact area moves across the particles' surfaces as they roll. The particles "grab" each other at the leading edge of the moving contact area, but micro-slip occurs along the trailing edge. As a result, the contact area migrates slightly faster along one particle than it does along the other: as the particles roll their incremental rolling traversal $d\mathbf{u}^{\text{roll}}$ is accompanied by a small amount of micro-slip and by a small incremental tangential sliding displacement $d\mathbf{u}^{\text{rigid,t}}$ (see equations [2.2], [2.3] and [2.13]). The relative amount of micro-slip displacement and rolling traversal is called the *creepage* and depends on the tangential force between the particles [KAL 67, JOH 85]:

$$\xi_{\text{s-s}} = \frac{|d\mathbf{u}^{\text{rigid,t}}|}{|d\mathbf{u}^{\text{roll}}|} = \mathcal{F}_{\text{s-s}}\left(\frac{|\mathbf{f}^{\text{t}}|}{\mu f^{\text{n}}}\right) \qquad [2.46]$$

where $\xi_{\text{s-s}}$ is the steady-state creepage that applies to steady rates of micro-slip displacement and rolling and $\mathcal{F}_{\text{s-s}}$ is the monotonic increasing creepage function. Creep-friction causes the two rollers in Figure 2.8 to rotate at slightly different rates. For the traditional problem of rail–wheel rolling interaction, the creepage is quite small (a locomotive wheel rim will typically move faster than the locomotive by only a fraction of a percent). The size of the micro-slip area in Figure 2.10(b) increases with increasing shearing traction between the two particles and covers the full contact area when the friction limit of equation [2.34] is reached: as a locomotive moves up a steepening grade, the micro-slip increases at the expense of forward advance, and the train eventually stalls, with all slip and no rolling.

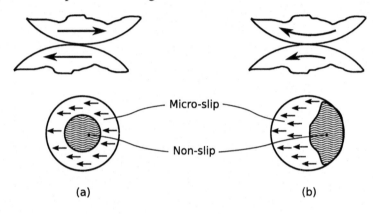

Figure 2.10. *Particle movements (top) and regions of micro-slip and non-slip within the circular contact area of a sphere–sphere contact (bottom) for two types of contact interaction: (a) Cattaneo-Mindlin tangential shearing; (b) creep-friction [KUH 14b]*

Kalker [KAL 00] derived the steady-state creepage function for sphere–sphere contact, expressed as:

$$\mathcal{F}_{s\text{-}s}\left(\frac{|\mathbf{f}^{n}|}{\mu f^{t}}\right) = \frac{8\mu}{(1-v)C_{11}} \frac{a}{R}\left[1 - \left(1 - \frac{|\mathbf{f}^{t}|}{\mu f^{n}}\right)^{1/3}\right] \qquad [2.47]$$

where R is the spheres' radius, a is the radius of the contact area (equation [2.23]), μ is the friction coefficient and C_{11} is a coefficient that depends on the spheres' Poisson ratio and ranges from 3.4 to 5.2 for ratios v of 0 and 0.5.

The author proposed a means of incorporating creep-friction effects in the analysis or simulation of granular materials, by incorporating a creepage function into the Cattaneo-Mindlin stiffness relation between tangential force and contact displacement [KUH 14b]. The incremental stiffness $\mathcal{K}_{\text{C-M}}(\cdot)$ follows the principles of section 2.3 and relates the increment of tangential force, $d\mathbf{f}^{\text{t}}$, to a given tangential displacement $d\mathbf{u}^{\text{rigid,t}}$ as in equation [2.35] (e.g. the Jäger algorithm [KUH 11] yields this force increment). The force increment, including the effects of both displacement and rolling, is approximated as:

$$d\mathbf{f}^{\text{t}} \approx \mathcal{K}_{\text{C-M}}\left(d\mathbf{u}^{\text{rigid,t}} - \frac{\mathbf{f}^{\text{t}}}{|\mathbf{f}^{\text{t}}|}|d\mathbf{u}^{\text{roll}}|\mathcal{F}_{\text{s-s}}\left(\frac{|\mathbf{f}^{\text{t}}|}{\mu f^{\text{n}}}\right)\right) \qquad [2.48]$$

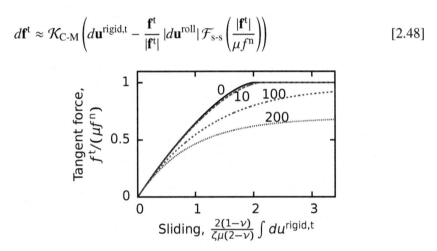

Figure 2.11. *The effect of creep-friction for two spheres that are initially pressed together and then undergo simultaneous displacement and rolling. Various ratios of rolling and displacement rates, $du^{\text{roll}}/du^{\text{rigid,t}}$, are designated with the displayed numbers. The results are based on equation [2.48] and Cattaneo-Mindlin theory, as expounded by Jäger [JÄG 05, KUH 11, KUH 14b]*

Figure 2.11 shows the effect of creep-friction on the tangential force between two spheres. The spheres are first pressed together to produce indentation ζ and then undergo a prolonged but steady combination of displacement and rolling motions. In the figure, the ratio $du^{\text{roll}}/du^{\text{rigid,t}}$ (as numbered) is assumed sustained with the ratio being the inverse of the creepage in equation [2.46]. The accumulating displacement (the sliding distance $\int du^{\text{rigid,t}}$) causes an increase in the tangential force. Note that a ratio of zero corresponds to the Cattaneo-Mindlin condition of pure tangential displacement of an initially indented contact (see Figure 2.5(d) with $du^{\text{rigid,t}} = 2\xi$). The effect of creep-friction is seen to be small, unless the rolling movement is many times greater than the rigid displacement. The reduction in the tangent force that occurs when rolling is concurrent with displacement requires a relatively large rolling distance, such that the two particle "roll past" the original contact patch of radius a.

3

Numerical Simulation

3.1. Numerical methods for discrete particle systems

Numerical models of discrete systems have become an important means of simulating and exploring the micro-scale behavior of granular systems, discerning the factors that affect macro-scale behavior and determining the micro-scale origins of the macro-scale response. Numerical simulations treat the particles as discrete objects that interact with neighboring particles through their contacts, through long-range interactions and through the pressure distributions within the pore fluid. These interactions cause each particle to move and rotate within its neighborhood of particles. Most methods allow finite displacements of the grains, allow particles to separate at their contacts or to freshly engage at new contacts, and include means of tracking particle motions as the assembly boundaries are displaced or loaded. Because granular materials are represented with individual grains, these methods differ from continuum models such as finite element (FEM) or boundary element models (BEM), which require a general and well-defined continuum description of the constitutive behavior at representative infinitesimal points. No such continuum description is required with discrete numerical models, as the macro-scale behavior is entirely derived from the discrete interactions of particles at their contacts. Numerical methods also enable an understanding of micro-scale mechanisms and their meso-scale evolution during loading, with the promise of determining and developing continuum descriptions of granular materials that are based on these mechanisms.

Discrete numerical models are applied to two general classes of problems: element simulations and boundary value problems. With an element simulation (or element test), the system of particles is intended as a surrogate of a continuum point, so that a material's continuum behavior and its underlying origins are derived from the discrete, grain-scale model. A relatively small numerical specimen with simple, uniform boundaries is subjected to various loading schemes, so that stiffness, strength and fabric are determined from this idealized material sample. In an element test, the numerical specimen serves as a representative volume element (RVE), in a role similar to that of a small physical, laboratory soil specimen, for which the simulation (or testing) results are presumed to represent the continuum response. Element simulations can also be used in a two-scale program of analyzing a large granular region: the discrete simulation determines material behavior at a small scale and informs a continuum description of soil behavior at the larger scale of a foundation system, earth slope or earth retention system. The number of particles in the element simulation is a measure of its numerical "size": a larger number of particles usually results in a better, more statistically representative simulation of material behavior but requires more computer time, memory and storage. Localized, meso-scale deformations (e.g. shear bands) are more likely within a large specimen, so we must decide beforehand whether such localization is to be allowed (even encouraged) or suppressed.

Although laboratory testing should be the ultimate arbiter of material behavior, numerical element simulations offer a number of advantages over laboratory testing. Among these advantages are the following:

1) Simulations provide insight into the micro-scale origins of granular behavior and an understanding of the underlying source of granular phenomena. These micro-scale aspects, besides being of interest in themselves, can aid in developing, testing and improving continuum models that are based on valid micro-mechanics.

2) Because discrete simulations are based on the particle shape, stiffness, and strength and on the interactions of particles at their contacts, they can reveal the effect of altering these micro-scale properties on the macro-scale behavior.

3) Numerical simulations can subject a granular element to a wide range of deformation conditions. Some three-dimensional codes permit the arbitrary

control of simple parallelepiped assemblies, by specifying a collection of any six components of the stress tensor, strain tensor or linear combinations of stress and strain. Arbitrary sequences of such loading can be stipulated in a simulation program to fully explore the history-dependent response of a material. These loading paths might require multiple sample preparations or even different testing apparatus in a physical laboratory.

4) After creating and loading a granular assembly to a given state, the same state and all of its micro-data (particle positions, contact forces, contact loading history, etc.) can be saved and reused as the reference state for any subsequent loading probe or sequence. In this manner, the same specimen can be probed or loaded with assurance that the subsequent behavior is entirely due to the loading direction and not to subtle differences in the initial sample.

5) Numerical techniques can be used for separating strain into its elastic and unrecoverable components, separating the stress-work into its separate contributions (changes in elastic energy, energy dissipation, etc.), determining the effects of non-homogeneous deformation patterns (patterns that include spatial gradients of strain or rotation) on stress evolution, discerning conditions that mark the incipient loss of material stability and uniqueness, etc.

The potential of element simulations is quite broad, and much of our current understanding of granular materials and their unusual characteristics (dilatancy, history dependence, internal force transmission, etc.) has been learned through simulation element tests.

Many difficult boundary value problems have also been addressed with discrete numerical simulations, including problems in the areas of mining, material transport and processing, foundation support, slope stability, *in situ* testing, tunneling and blasting. When modeling a full boundary value problem, the simulation must faithfully represent the walls, the free boundaries, the self-weight of the material and other aspects of the problem that is being simulated. Because of the large numbers of particles that are involved in simulating a full boundary value problem, these simulations are computationally intensive and have driven novel advances in numerical algorithms and more effective use of computer hardware (parallelization, use of multi-core graphical processors, etc.). The number of particles in most full-scale geotechnical problems is usually much larger than can be explicitly modeled with current computational resources, but a grain-for-grain simulation might not be required. Larger, surrogate grains can be used in the simulation of a boundary value problem, provided that the particles' aggregate, meso-scale behavior and self-weight are similar to that of the targeted soil.

Before conducting a discrete simulation, we must honestly assess the intention of the simulation, as the methods used and the complexity of their execution will depend on this intent. For example, we should ask whether the purpose is to gain a general understanding of a conceptual problem or whether we intend to model a specific soil or granular material with a fair level of fidelity to laboratory or field tests. At present, the latter purpose is much less common, as effective and reliable models are only now being developed and the inter-related effects of various modeling aspects – particle shape, contact behavior and assembly generation – are still being explored. Simulating a specific material can require great effort to calibrate a model that reliably reproduces laboratory results.

We should also decide whether the particles' inertias are an essential aspect of the problem being modeled and whether the full dynamics of a situation must be captured (for example, when modeling rapid flows with sparse densities, sedimentation problems, landslide run-out problems, etc.). In these dynamical problems, particle interactions can be brief and collisional, and greater attention is required in modeling the rotational tumbling and twirling of the particles and in modeling their collisional interactions. At the other extreme are *quasi-static* simulations which are intended to model problems in which the loading is slow and particle inertias are of secondary concern. Maintaining near-equilibrium conditions among the particles is the primary computational goal of quasi-static simulations, and greater attention is required in formulating a robust contact model for sustained, persistent particle interaction.

Several classification schemes have been proposed for the various discrete simulation methods that have been applied to granular materials [CUN 89a, BAR 98]. Among the distinguishing characteristics of each method are its treatment of the particles at their contacts (soft, deformable contacts or hard, non-penetrating contacts), its treatment of the particle bodies (rigid or deformable) and its numerical approach to satisfying the equilibrium (or dynamics) of all particles (as a global solution or through an iteration of local solutions). Three common methods that encompass the range of these options are presented in separate sections:

– the discrete element method (DEM): this method, originally called the distinct element method, uses a deformable contact model with rigid particles and applies dynamic relaxation to approximately solve the local equations of motions (section 3.2);

– stiffness matrix methods: these methods are derived from principles of matrix structural analysis and include the discontinuous deformation analysis (DDA) method. The contacts and the particle bodies can be rigid or deformable, but these methods attempt a global solution to the equations of motion (section 3.3);

– the contact dynamics (CD) method: this method was originally intended for rigid particles with hard contacts and applies an implicit integration scheme to approximately solve the local equations of motions of the particles (section 3.4).

The three methods are presented separately in sections 3.2–3.4. Each of these methods was pioneered in the 1970s or 1980s and has undergone considerable refinement. Each method has evolved from its early application to simple two-dimensional shapes (disks or rectangles), simple particle interactions, and basic boundary conditions to more complex particle geometries, interactions and boundaries. The DEM has the longest history and has been extended to the widest variety of conditions, including multi-phase materials, long-range particle interactions, breakable particles and close integration with finite element methods.

3.1.1. *Particle shape*

Discrete particle simulations require an unambiguous geometric description of the particles' sizes, shapes, locations and orientations. Particle shapes can be classified as two or three dimensional, as smooth or with edges and corners, as convex or non-convex and as rotund or elongated. Because they are smooth, convex and rotund and have a simple geometric representation, disks and spheres have been the predominant choice for code development and for simple simulations: a single parameter (radius) is required to describe the shape, contact detection and resolution are simple, and orientation is irrelevant (Figure 3.1(a)). For two-dimensional problems, ellipses are only slightly more complex, requiring two shape parameters (the semi-axes) and a single parameter to describe orientation (Figure 3.1(a)).

Contact detection between two ellipses involves the solution of a quadratic equation [TIN 91, ROT 91]. The three-dimensional counterpart of an ellipse, an ellipsoid, can be of two types: simple spheroids with two distinct semi-axes (a solid of revolution generated by an ellipse) and general ellipsoids having three different semi-axes. The former includes elongated (prolate) spheroids and squat (oblate) spheroids, and their orientation can be simply described with two Euler angles. As with all non-spherical and non-symmetric three-dimensional shapes, a general ellipsoid with three distinct semi-axes requires the use of three Euler angles, a rotation matrix or a quaternion to specify its orientation (section 3.1.6). Resolving the contact of two ellipsoids involves solution of a quartic equation [LIN 97, HOP 04]. Moreover, the exact rotational kinetics of a three-dimensional body, although simple for spheres and other symmetric shapes, is computationally demanding for general shapes such a ellipsoids (section 3.2.3). The kinetics of spheroids and other shapes having an axis of symmetry is of intermediate complexity [LIN 97].

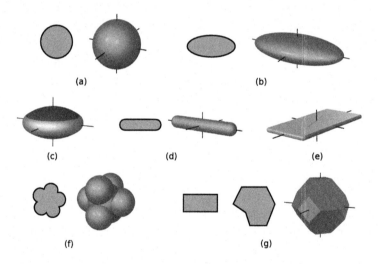

Figure 3.1. *Idealized particle shapes for simulations: (a) disk and sphere; (b) ellipse and ellipsoid; (c) ovoid; (d) dilated lines; (e) dilated polygon; (f) disk- and sphere-clusters; (g) polygons and a polyhedron.*

Other shapes can be formed as composites of simple shapes, such as spheres, tori, ellipsoids and cylinders. The composite of a torus and two spheres (an ovoid) approximates a spheroid (either prolate or oblate) but with simpler contact detection (Figure 3.1(c)) [KUH 03b]. Hopkins [HOP 04] proposed the use of dilated shapes (also called spheropolygons [ALO 08]), formed from a basic shape (a point, set of points, line, polygon, etc.) by placing the center of a sphere (or disk in 2D) at every point within the basic shape (i.e. a morphologic dilation). For example, a three-dimensional line becomes a cylindrical rod with spherical caps (spherocylinder, or in 2D, a rectangle with circular caps, Figure 3.1(d)). Contact detection is made straightforward by finding the points along the two basic shapes that have the minimum separation, using a "rubber band" algorithm to minimize this distance. Boton *et al.* [BOT 13] have used dilated polygons to form three-dimensional smooth platy shapes that resemble clay particles (Figure 3.1(e)).

By using a set of points as the basic shape, complex dilated non-convex shapes can be formed as composite clusters of overlapping disks or spheres (Figure 3.1(f)) [THO 99, VUQ 00, FAV 01]. The shapes of real gravel and sand particles can be closely approximated by using overlapping spheres of different radii [FER 10]. Contact detection between these sphere-clusters or disk-clusters is as simple as finding sphere–sphere (or circle–circle) distances between the sphere centers of the two particles. Contact characterization is also greatly simplified, as all the contacts are between smooth spherical surfaces. As with other non-sphere shapes, the orientation of a clustered particle must be tracked with an orientation matrix or quaternion, so that each sphere in the cluster can be located relative to the particle's center (section 3.1.6). Because the shapes are non-convex, a pair of particles can touch each other at more than one contact, which requires more involved data structures to store the lists of potential contacts (section 3.1.5).

Complex shapes can also be created with polygons and polyhedra. By increasing the number of faces, a polyhedron's shape can closely approximate that of realistic sand and gravel particles (Figure 3.1(g)) [CUN 88b, AZÉ 07, AZÉ 09]. The elegant common-plane concept of Peter Cundall [CUN 88b] is commonly used to reduce the contact detection time for convex shapes: rather than testing the proximity of each face, edge and vertex of one particle with those of a candidate particle (a quite lengthy process), we test the proximity of each vertex of one particle with an imaginary common plane that is shifted and reoriented to remain midway between the two particles. A contact is established when both particles touch the common plane.

Andrade and his co-workers [AND 12] have created quite complex shapes using non-uniform rational basis splines (NURBS). These smooth shapes, commonly used in computer graphics and animation, can be non-convex and can closely approximate the shapes of digitized particle images. The shape is formed from a smooth (infinitely differentiable) piecewise polynomial closed surface defined by non-uniformly spaced control points. An important property of these shapes is that the polyhedral convex hull of the control points forms the convex hull of the entire shape, so that the proximity of two NURBS particles can be established from the separation (or contact) of their polyhedral hulls. An iterative procedure must be used in the final contact determination, and once a contact is verified, the overlap, contact normal and surface curvatures can be found.

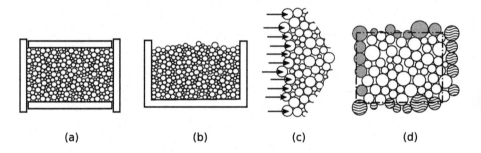

Figure 3.2. *Boundaries for discrete numerical simulations: (a) rigid boundaries on all sides; (b) free top boundary; (c) flexible (force) boundary on left side; (d) periodic boundaries on all sides*

3.1.2. *Boundaries*

The boundaries of a particle assembly can be free, rigid, flexible, periodic or combinations of these types (Figure 3.2). Rigid boundaries are modeled as separate discrete particles that interact with the interior particles, and the movement (or stationarity) of the a wall is independently controlled rather than reacting to the forces imparted by interior particles. A rigid boundary can be flat or curved, smooth or serrated, or a rigidly moving layer of particles. Contacts between interior particles and a rigid boundary are treated in the same manner as those between the interior particles, but distinct stiffness and frictional characteristics can be assigned at the boundary. Because a rigid boundary can be treated as a single large particle, the average stress along the boundary can be controlled by applying and maintaining a desired external force $\mathbf{b}^p$ on the particle (see Figure 3.4). Although rigid boundaries are easily coded, the corners between flat walls create local dead zones that restrict particle motions, which may be undesirable when more homogeneous conditions are sought. Rigid walls can also influence, or even suppress, the formation of shear bands or other localized deformations that might otherwise form near the periphery of an assembly.

Flexible boundaries control the forces on peripheral particles, imparting a controlled stress within the immediate interior of an assembly while allowing greater freedom of movement among the peripheral particles. These boundaries are an analog of the flexible rubber membranes used in soil testing. Unlike rigid walls, flexible boundaries permit interior shear bands to extend to the boundary, possibly folding or kinking the boundary. Although rubber membranes apply a normal isotropic stress to the peripheral particles, a simulated flexible boundary can be configured to transmit both normal and shear stress and either isotropic or anisotropic (deviatoric) normal stress around the assembly's periphery. For two-dimensional assemblies, the boundary can consist of straight "links" connecting the centers of peripheral particles, exerting a controlled force to the particles [BAR 92b]; in three dimensions, virtual triangular plates, corresponding to a Delaunay tessellation of the peripheral particles, can be connected to the peripheral particles' centers [KUH 95, KRE 01]. A similar technique is to sheath the interior assembly within a layer of outer particles that serve as the flexible boundary. Exterior forces are applied to these outer particles, which then transfer their forces to the assembly's interior [CHE 08].

Periodic boundaries are an effective and simple means of containing an assembly when conditions of relatively uniform stress and density are desired [CUN 88a], and, as such, periodic boundaries are the most commonly used boundary type for element tests (Figure 3.2(d)). Particles along a periodic boundary interact with particles along the opposite boundary and are allowed to cross a boundary. When a particle passes over a periodic boundary, an "image" particle emerges from the opposite boundary to interact with particles on both (opposing) sides of the assembly. In three dimensions, a box of six periodic boundaries can be defined with an upper-triangular matrix $\mathbf{A}$, which is the product of the deformation gradient $\mathbf{F}$ (equation [1.86]) and the matrix of original cell dimensions $\mathbf{X}^{\text{cell}}$, such that $\mathbf{A} = \mathbf{F} \cdot \mathbf{X}^{\text{cell}}$. The separation vector between two points, $\mathbf{x}^p$ and $\mathbf{x}^q$, and the relative movement,

$d\mathbf{u}^q - d\mathbf{u}^p$, of the two points are computed as:

$$\mathbf{x}^q - \mathbf{x}^p \quad \leftarrow \quad (\mathbf{x}^q - \mathbf{x}^p) - \mathbf{A} \cdot \text{round}\left(\mathbf{A}^{-1} \cdot (\mathbf{x}^q - \mathbf{x}^p)\right) \qquad [3.1]$$

$$d\mathbf{u}^q - d\mathbf{u}^p \quad \leftarrow \quad (d\mathbf{u}^q - d\mathbf{u}^p) - d\mathbf{A} \cdot \text{round}\left(\mathbf{A}^{-1} \cdot (\mathbf{x}^q - \mathbf{x}^p)\right) \qquad [3.2]$$

where the round( ) function rounds its vector argument to the vector of nearest integers and $d\mathbf{A} = d\mathbf{F} \cdot \mathbf{X}^{\text{cell}}$ (i.e. $dA_{ij} = dF_{ik}X_{kj}^{\text{cell}}$, where $d\mathbf{F}$ is related to the instantaneous strain rate with equation [1.93]). Matrix $\mathbf{X}^{\text{cell}}$ is diagonal for an initial rectangular box but contains off-diagonal terms with a skewed box. Equations [3.1] and [3.2] can be used in detecting contacts (section 3.1.5) and in computing the contact movements (equations [2.1]–[2.4]). Periodic boundaries, however, do influence, and can impede or even prevent, the formation of shear bands that might otherwise fully pass through an assembly, since a nearly planar shear band that begins at one boundary must "wrap" across the assembly to meet (and coincide with) itself at the opposite boundary. By recognizing and taking advantage of this geometric precondition and by using an unusual arrangement of periodic boundaries, Sun *et al.* [SUN 13] were able to study the interior of a shear band that wrapped across periodic boundaries. Unlike rigid and flexible boundaries, which permit direct application of boundary forces and stress, a servo-control algorithm is required to control the average stress of an assembly that is contained within periodic boundaries. The algorithm must continually adjust the assembly dimensions (i.e. the matrix $\mathbf{A}$), so that the desired stress components $\overline{\sigma}_{ij}$ in equation [1.53] are faithfully maintained at their desired, target values. The author introduced an adaptive servo-algorithm for element tests that allows control of an arbitrary set of stress components [KUH 92].

3.1.3. *Assembly and particle sizes*

Computational feasibility and effort depend on the number of particles, so we must carefully consider the assembly size required for a suitable simulation. The actual size of most geotechnical problems involves billions or trillions of soil particles within the problem domain, which is certainly outside the limits of current (and future) computer capabilities. A reasonable compromise can be to increase the particle size relative to the domain size, knowing that each 10-fold increase in particle size reduces the number of particles and the computational hardware and time by a factor of a thousand. In this manner, the larger numerical particles, although not the same size as the actual particles, can be calibrated to capture the representative character of the actual material's behavior. The question of assembly size then rests on the number of these surrogate particles that are required to accurately represent a granular region undergoing deformation.

With laboratory strength testing of soils, the specimen dimensions are typically recommended as being at least 10 times the size of the largest particles, so that the specimen is a representative volume (RVE) of a possibly larger region of the same material. Further guidance on assembly size can be derived from sets of element simulations using assemblies that contain different numbers of particles but are subject to the same bulk deformation. The average strength of an assembly of particles typically decreases with an increase in the number of particles within the assembly, due to the less restrictive influence of the boundary particles. The author has found that with two-dimensional regions surrounded by periodic or rigid boundaries, this influence of assembly size does not appreciably affect assemblies with more than a thousand particles, even though strength is more variable (and less certain) with smaller assemblies [KUH 09]. By testing multiple assemblies of different sizes, the standard deviations of strengths among assemblies with 256 and 1000 particles were reduced from 12% to 5% of the mean stress. Although two-dimensional assemblies with a thousand or more particles can capture the material behavior under conditions of fairly homogeneous deformation, the thickness of a shear band is typically 10–15 particle widths, and a two-dimensional assembly with fewer than a thousand particles will suppress the emergence of these localization features. Three-dimensional simulations with sphere-clusters have shown that good reproducibility is attained with 8000 particles [SAL 09], although three-dimensional assemblies with over a hundred thousand particles are required to permit the emergence of shear bands [SUN 13].

We must also decide on the range and distribution of particle sizes before creating a granular assembly. Mono-disperse assemblies (single, uniform particle size) are problematic, as the particles tend to crystallize into closely packed clusters or regions, and deformation tends to occur as slips along crystallographic planes. On the other hand, for a given size of the largest particles, a wider range of particle sizes requires a greater number of particles to fill a region. To gain some perspective on this situation, consider the Unified Soil Classification System that is commonly used for classifying soils and mineral aggregates, for which a "well graded" sand has a coefficient of uniformity (C_u) greater than 6: the ratio of particle sizes that demark the smallest 60% and smallest 10% of a material by cumulative weight (section 1.1) [AST 06]. If an assembly is composed of only two sizes (bi-disperse), a factor of 6 in size corresponds to a factor of 6^3 in number, so that each large particle must have 256 smaller siblings. Representing a well-graded soil requires more particles than is reasonable for most simulations. As such, most numerical simulations are of "poorly graded" soils having a much smaller C_u, between 1.5 and 4.0.

3.1.4. *Assembly creation*

Besides establishing the particle shapes, particle sizes and assembly size, we must create an initial domain filled with particles. Creating the initial assembly is often the most laborious process of a simulation. This is certainly the case when the assembly is intended to represent a particular soil that has a specific initial density and must exhibit a strength and stiffness that closely resembles the target soil. Four techniques are commonly used for creating assemblies. The first method mimics natural processes, by pluviating (raining) particles into a container with rigid boundaries, producing an assembly with an initial, inherent anisotropy imparted by the gravitational direction. The density of the resulting assembly can be changed by adjusting the gravitational force, the viscosity of the virtual fluid through which the particles fall and the contact friction coefficient μ of the contacts. Slower, more sluggish pluviation produces assemblies of lower density (higher void ratio e); whereas, reducing μ increases density and increases the average coordination number of the initial packing [SIL 02].

Constructive methods place particles one-by-one into an assembly, intentionally nestling each new particle among the existing neighbors in a progressively growing assembly [FEN 01, BAG 05]. These methods require an algorithm to create a geometric fit and can be difficult to adjust to attain a specific density and packing anisotropy. The space that remains to accommodate the final few particles usually results in a small loose pocket within the denser assembly.

A common method for creating isotropic assemblies is to "grow" (dilate) the particles from initial seed points, allowing the particles to shift and accommodate their neighbors until they eventually attain a target criterion.

Possible criteria include filling the container to arrive at a target mean stress or attaining a predetermined particle size [ITA 14]. To create an assembly with a target density, the contact friction coefficient μ can be adjusted at the start of (or during) the particle dilation process [SAL 09]. Multiple trials are usually required to create an assembly with the target density, particle size and mean stress.

In a variation of a method proposed by Cundall [CUN 79], the author usually begins with a sparse arrangement of seed particles, each with its final size, and then slowly and isotropically contracts the boundaries (either periodic or rigid), allowing the particles to shift and adjust their positions while maintaining their original sizes, until the entire assembly "seizes" with a target mean stress. Greater densities are created in stages by perturbing the particles of the previous assembly (by assigning random velocities) and then further collapsing the boundaries until the assembly seizes again. A wide range of densities can be created by adjusting the friction coefficient, the perturbing velocities and the number of densifying stages [THO 00]. Anisotropic assemblies can be created by contracting the surrounding boundaries at different rates in the three orthogonal directions.

3.1.5. *Contact detection*

One of the most computationally demanding processes during a simulation is determining which pairs of particles are touching (or are potentially touching) and then characterizing each contact with respect to its overlap and the surfaces' shapes at their contact. A naive search among all $N(N-1)$ pairs of an N-particle assembly is impractical except for small assemblies. A common approach is to create a rectangular grid that covers the assembly and partition it into a system of bins which are somewhat larger than the largest particle. The particles are periodically assigned to their particular bins, so that contact detection is restricted to those particle pairs that lie within the same or adjacent bins. In this manner, the number of particle pairs that are queried for contact detection is proportional to the number of particles N rather than N^2. Another N-proportional strategy is to place each particle within a circumscribing boxes (or rectangular box) with faces that are parallel to those of the boxes of other particles. The boxes, which move with their particles, can then be projected onto the coordinate axes so that overlapping boxes are identified [MUT 07]. Contact detection can then proceed among the particles within these overlapping circumscribing cubes.

Because the contact search process requires considerable effort if done with each time step, the search process can instead be used to produce a list of potential candidate pairs of particles. Subsequent contact detection is then restricted to this near-neighbor list (Verlet list) during the next several time steps. Although it greatly reduces the frequency of a full search for potential contacts, the use of near-neighbor lists demands a well-defined criterion for designating potential contacts (a minimum separation criterion) and a robust means of determining the necessity of updating the near-neighbor list so that pairs of rapidly moving particles, not already in the list, do not come into contact before the list is updated [CUN 88b].

A near-neighbor scheme inevitably requires data structures for storing lists of potential contacts. Linked lists are an efficient means for storing and retrieving this information from one time step to the next [KNU 73]. For example, the association of each bin with its adjacent bins can be stored within the linear structure of a linked list. The near-neighbor list of particle pairs can also be stored as a linked list to facilitate rapid look-up and updating. Contacts can be added and removed from a linked list with a few lines of code, and storage space that was originally used for contact information (overlaps, contact history, etc.) but set aside (discarded) for future usage can be referenced with a heap data structure to efficiently conserve computer storage. With composite particles of multiple elements (for example, Figure 3.1(f)), the lists of potentially contacting particle pairs must point to other lists of potentially contacting elements within each of the two particles. Models of contact stiffness and force are usually history dependent (section 2.3), and once a potential contact is verified as an actual contact, the contact's history must be also be stored, most efficiently with another linked list. The maintenance of these inter-linked data lists is an inevitable constituent of most simulation codes.

3.1.6. *Particle orientation and rotation kinematics*

Numerical simulations with non-spherical particles must track the orientations and rotations of the grains, locate the contact points between neighboring grains, and compute the directions that are normal and tangential to the grains' surfaces at the contacts. Once these tasks are accomplished, the methods in section 2.1 can be applied to compute the components of displacement, rolling and twisting at the contact. The orientation of a non-spherical particle can be expressed in a number of ways, the most common being with Euler angles, rotation matrices and orientation quaternions. The latter two methods are most appropriate for particle simulations, since they relate most directly to the rotational velocity of a particle. With both methods, we must distinguish between two coordinate frames. The body "l" frame (local frame, element frame) is attached to, and moves and rotates with, the particle. This frame is affixed to the reference point χ^p of a particle p (most conveniently at its centroid), and its three Cartesian directions are usually aligned with the particle's principal axes of inertia: the frame $\{\chi^p, \mathbf{e}_1^{l,p}, \mathbf{e}_2^{l,p}, \mathbf{e}_3^{l,p}\}$ in Figure 3.3. The global "g" frame (space frame, inertial frame) is attached to origin O and is either fixed or moves in rectilinear motion at a constant rate but does not rotate: the frame $\{O, \mathbf{e}_1^g, \mathbf{e}_2^g, \mathbf{e}_3^g\}$.

The global position $\mathbf{x}^{g,\bullet}$ of a point "$\bullet$" within or on a particle p can be computed from the location $\mathbf{x}_0^{g,p}$ of the particle's centroid χ^p relative to the global frame, the location $\mathbf{x}^{l,\bullet}$ of the point with respect to the particle's body frame, and the particle's orientation matrix (rotation matrix or metric tensor) $\mathbf{R}^p$:

$$\mathbf{x}^{g,\bullet} = \mathbf{R}^p \cdot \mathbf{x}^{l,\bullet} + \mathbf{x}_0^{g,p} \qquad [3.3]$$

or $x_i^{g,\bullet} = R_{ij}^p x_j^{l,\bullet} + x_{0,j}^{g,p}$, where the superscripts "g" and "l" designate coordinates in the global and local frames, respectively (Figure 3.3). Position $\mathbf{x}^{l,\bullet}$ is relative to the particle's centroid (i.e. $\mathbf{x}^{l,\bullet} = 0$ at the centroid). Column j of matrix $\mathbf{R}^p$ is the projection of the unit vector $\mathbf{e}_j^{l,p}$ of the j-th direction in the local frame onto the three $\mathbf{e}_i^g$ coordinate directions of the global frame:

$$R_{ij}^p = \mathbf{e}_i^g \cdot \mathbf{e}_j^{l,p} \qquad [3.4]$$

Matrix $\mathbf{R}^p$ is an orthogonal matrix, meaning that its columns (and its rows) are linearly independent and orthogonal, such that its inverse is simply its transpose:

$$(\mathbf{R}^p)^{-1} = (\mathbf{R}^p)^T \quad \text{and} \quad \mathbf{R}^p \cdot (\mathbf{R}^p)^T = (\mathbf{R}^p)^T \cdot \mathbf{R}^p = \mathbf{I}_{\bar{3}} \qquad [3.5]$$

where $\mathbf{I}_{\bar{3}} = \delta_{ij}$ is the identity matrix. The rotation matrix will change with the particle's orientation during a numerical simulation, but the positions of points within the particle, as expressed in their local coordinates $\mathbf{x}^{l,\bullet}$, remain unchanged. For example, the location of important points, such as the vertices of a polyhedral particle, can be established at the beginning of a simulation, and equation [3.3] is used thereafter to locate these points within the inertial frame.

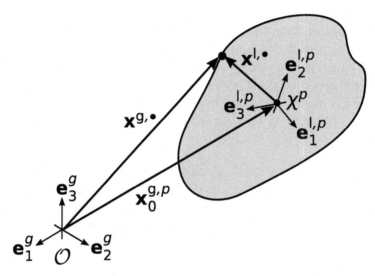

Figure 3.3. *Global and local reference frames of a particle p*

The inertia tensor $\mathbf{I}$ (which should not be confused with the identity matrix $\mathbf{I}_{\bar{3}}$) is used to compute the particle's dynamics, and its nine values are usually determined in the local frame at the start of a simulation and then updated in the inertial frame as the simulation progresses. The tensor $\mathbf{I}^{l,p}$ relative to the body frame of particle p and taken about the centroid of the particle is defined as:

$$\mathbf{I}^{l,p} = \int_{\Omega^p} \rho(\mathbf{x}^{l,\bullet})[(\mathbf{x}^{l,\bullet} \cdot \mathbf{x}^{l,\bullet})\mathbf{I}_{\bar{3}} - \mathbf{x}^{l,\bullet} \otimes \mathbf{x}^{l,\bullet}] \, dV \qquad [3.6]$$

$$= \begin{bmatrix} I_{22}^{l,p} + I_{33}^{l,p} & -I_{12}^{l,p} & -I_{13}^{l,p} \\ -I_{12}^{l,p} & I_{11}^{l,p} + I_{33}^{l,p} & -I_{23}^{l,p} \\ -I_{13}^{l,p} & -I_{23}^{l,p} & I_{11}^{l,p} + I_{22}^{l,p} \end{bmatrix} \qquad [3.7]$$

where the moment of inertia $I_{ij}^{1,p}$ is a weighted integration of dyads $\mathbf{x}^{1,\bullet} \otimes \mathbf{x}^{1,\bullet} = x_i^{1,\bullet} x_j^{1,\bullet}$, as:

$$I_{ij}^{1,p} = \int_{\Omega^p} \rho(\mathbf{x}^{1,\bullet}) \, x_i^{1,\bullet} x_j^{1,\bullet} \, dV \tag{3.8}$$

and $\rho(\mathbf{x}^{1,\bullet})$ is the mass density at point $\mathbf{x}^{1,\bullet}$ within the particle. The inertia tensor $\mathbf{I}^{1,p}$ in the body frame, which remains constant, is related to its expression in the global (inertial) frame through the tensor transformation:

$$\mathbf{I}^{g,p} = \mathbf{R}^p \cdot \mathbf{I}^{1,p} \cdot (\mathbf{R}^p)^{-1} \quad \text{or} \quad I_{ij}^{g,p} = R_{ik}^p I_{kl}^{1,p} R_{jl}^p \tag{3.9}$$

By using equations [3.6]–[3.9], the particle's inertia tensor is computed once in the simpler body frame, and thereafter is easily rotated to obtain its components in the global frame. The most convenient representation of the inertia tensor is found by aligning the body frame axes with the principal inertia axes, for which the tensor is:

$$\left(\mathbf{I}^{1,p}\right)_{\text{principal}} = \begin{bmatrix} I_1^{1,p} & 0 & 0 \\ 0 & I_2^{1,p} & 0 \\ 0 & 0 & I_3^{1,p} \end{bmatrix} \tag{3.10}$$

with $I_i^{1,p}$ being the moment of inertia about the ith principal axis.

As a particle moves and rotates, an inertial (global) observer will note changes in the global coordinates $\mathbf{x}^{g,\bullet}$ of a point "$\bullet$" within the particle, due to changes in the particle's position $\mathbf{x}_0^{g,p}$ (i.e. translations) and changes in the particle's rotation matrix $\mathbf{R}^p$. The rotational velocity vector $\omega^{g,p}$ of particle p about an axis aligned with $\omega^{g,p}$, as viewed by a global observer, will change the elements of matrix $\mathbf{R}^p$ at the rate:

$$\dot{\mathbf{R}}^p = \omega^{g,p} * \mathbf{R}^p = \begin{bmatrix} 0 & -\omega_3^{g,p} & \omega_2^{g,p} \\ \omega_3^{g,p} & 0 & -\omega_1^{g,p} \\ -\omega_2^{g,p} & \omega_1^{g,p} & 0 \end{bmatrix} \begin{bmatrix} R_{11}^p & R_{12}^p & R_{13}^p \\ R_{21}^p & R_{22}^p & R_{23}^p \\ R_{31}^p & R_{32}^p & R_{33}^p \end{bmatrix} \tag{3.11}$$

where $\omega^{g,p}$ has components:

$$\omega^{g,p} = \begin{bmatrix} \omega_1^{g,p} \\ \omega_2^{g,p} \\ \omega_3^{g,p} \end{bmatrix} \qquad [3.12]$$

and the "$*$" operator (between the vector $\omega^{g,p}$ and the matrix $\mathbf{R}^p$) is defined in equation [3.11]. The incrementally altered rotation matrix, $\mathbf{R}^p + d\mathbf{R}^p$, is computed by using the particle's rotation increment $d\theta^{g,p} = \omega^{g,p} \, dt$ in place of $\omega^{g,p}$ in equations [3.11] and [3.12].

Although rotation matrices greatly simplify the shifting back and forth between body and inertial frames, an unfortunate difficulty arises when particle rotations require frequent updating of $\mathbf{R}^p$ (perhaps with each time step) by repeated application of equation [3.11]. Each application of this equation with a small but finite $d\theta^{g,p}$ can break the orthogonality condition of equation [3.5]. Restoring orthogonality requires a procedure such as the Gram–Schmidt process, which is computationally expensive. The use of quaternions makes this process much simpler, so that quaternions are often used in tandem with rotation matrices.

Quaternions offer a compact and efficient means of expressing a particle's orientation and for shifting between coordinate frames. They are commonly used in simulation codes to store particle orientations and to affect particle rotations. The orientation of particle p is expressed with a four-component unit quaternion $\mathring{\mathbf{q}}^p$, expressed as:

$$\mathring{\mathbf{q}}^p = \langle q_0^p, \mathbf{q}^p \rangle = \langle q_0^p, q_1^p, q_2^p, q_3^p \rangle \qquad [3.13]$$

in which q_0^p and $\mathbf{q}^p$ are the real and imaginary parts of $\mathring{\mathbf{q}}^p$. A quaternion is a *unit* quaternion when it has unit magnitude, $(q_0^p)^2 + (q_1^p)^2 + (q_2^p)^2 + (q_3^p)^2 = 1$. A unit quaternion represents a rotation in the following sense. A vector $\mathbf{x}^{1,\bullet}$ in the local frame is represented by the four-component quaternion $\mathring{\mathbf{x}}^{1,\bullet} = \langle 0, \mathbf{x}^{1,\bullet} \rangle$, and the coordinates of this vector relative to the global frame, centered at χ^p and expressed as the four-component quaternion $\mathring{\mathbf{x}}^{g,\bullet} = \langle 0, \mathbf{x}^{g,\bullet} \rangle$

$$\mathring{\mathbf{x}}^{g,\bullet} = \mathring{\mathbf{q}}^p \circ \mathring{\mathbf{x}}^{1,\bullet} \circ \mathring{\overline{\mathbf{q}}}^p \qquad [3.14]$$

In this expression, $\overset{\circ}{\bar{\mathbf{q}}}{}^p$ is the quaternion conjugate of $\overset{\circ}{\mathbf{q}}{}^p$, in which the imaginary part is reversed in sign, $\overset{\circ}{\bar{\mathbf{q}}}{}^p = \langle q_0^p, -\mathbf{q}^p \rangle$. The operator "$\circ$" in equation [3.14] designates the product of two quaternions, which can be computed as a matrix product:

$$\overset{\circ}{\mathbf{c}} = \overset{\circ}{\mathbf{a}} \circ \overset{\circ}{\mathbf{b}} \quad \Rightarrow \quad \begin{bmatrix} c_0 \\ c_1 \\ c_2 \\ c_3 \end{bmatrix} = \begin{bmatrix} a_0 & -a_1 & -a_2 & -a_3 \\ a_1 & a_0 & -a_3 & a_2 \\ a_2 & a_3 & a_0 & -a_1 \\ a_3 & -a_2 & a_1 & a_0 \end{bmatrix} \begin{bmatrix} b_0 \\ b_1 \\ b_2 \\ b_3 \end{bmatrix} \qquad [3.15]$$

Any unit quaternion $\overset{\circ}{\mathbf{q}}$ can be expressed in the form:

$$\overset{\circ}{\mathbf{q}} = \left\langle \cos\frac{\varphi}{2}, \left(\sin\frac{\varphi}{2}\right)\mathbf{n} \right\rangle \qquad [3.16]$$

where $\mathbf{n}$ is a unit vector. The quaternion expression in equation [3.14] rotates vector $\mathbf{x}^{1,\bullet}$ around the unit vector $\mathbf{n}$ by angle φ. A rotational transformation with unit quaternions, such as that in equation [3.14], can be written in matrix form as:

$$\overset{\circ}{\mathbf{a}}{}^g = \overset{\circ}{\mathbf{q}} \circ \overset{\circ}{\mathbf{a}}{}^l \circ \overset{\circ}{\bar{\mathbf{q}}} = 2\mathbf{Q} \cdot \mathbf{a} \qquad [3.17]$$

where the 3×3 matrix $\mathbf{Q}$ is related to rotation matrix $\mathbf{R}$ as:

$$\tfrac{1}{2}\mathbf{R} = \mathbf{Q} = \begin{bmatrix} \frac{1}{2} - q_2^2 - q_3^2 & q_0 q_3 + q_1 q_2 & q_1 q_3 - q_0 q_2 \\ q_1 q_2 - q_0 q_3 & \frac{1}{2} - q_1^2 - q_3^2 & q_0 q_1 + q_2 q_3 \\ q_1 q_3 + q_0 q_2 & q_2 q_3 - q_0 q_1 & \frac{1}{2} - q_1^2 - q_2^2 \end{bmatrix} \qquad [3.18]$$

Equations [3.3] and [3.14] rotate a vector from its body frame to the corresponding vector in the inertial frame. Among its other advantages, the quaternion rotation is embodied in only four numbers (those in $\overset{\circ}{\mathbf{q}}{}^p$) rather than the nine numbers of $\mathbf{R}^p$.

A particle's orientation quaternion $\overset{\circ}{\mathbf{q}}{}^p$ will change as the particle rotates within the global frame. Rotational velocity $\omega^{g,p}$, as viewed by a global observer, will change the elements of matrix $\overset{\circ}{\mathbf{q}}{}^p$ at the rate:

$$\frac{d\overset{\circ}{\mathbf{q}}{}^p}{dt} = \tfrac{1}{2}\overset{\circ}{\omega}{}^{g,p} \circ \overset{\circ}{\mathbf{q}}{}^p \qquad [3.19]$$

where $\overset{\circ}{\omega}{}^{g,p} = \langle 0, \omega^{g,p} \rangle$ and the quaternion product is expressed in matrix form as:

$$\tfrac{1}{2}\overset{\circ}{\omega} \circ \overset{\circ}{\mathbf{q}} = \frac{1}{2} \begin{bmatrix} 0 & -\omega_1 & -\omega_2 & -\omega_3 \\ \omega_1 & 0 & -\omega_3 & \omega_2 \\ \omega_2 & \omega_3 & 0 & -\omega_1 \\ \omega_3 & -\omega_2 & \omega_1 & 0 \end{bmatrix} \begin{bmatrix} q_0 \\ q_1 \\ q_2 \\ q_3 \end{bmatrix}$$

[3.20]

so that the updated quaternion after increment dt is $\overset{\circ}{\mathbf{q}}{}^p + d\overset{\circ}{\mathbf{q}}{}^p$. The quaternion rate in equation [3.19] is the complement of the rate of the rotation matrix in equation [3.11]. As with the rotation matrix, a quaternion will drift from the unit length condition after repeated updates, but restoring unit length is a simple matter of re-normalizing the quaternion: by dividing the quaternion by its magnitude, $\overset{\circ}{\mathbf{q}}{}^p \rightarrow \overset{\circ}{\mathbf{q}}{}^p / |\overset{\circ}{\mathbf{q}}{}^p|$, where magnitude $|\overset{\circ}{\mathbf{q}}{}^p| = \sqrt{(q_0^p)^2 + (q_1^p)^2 + (q_2^p)^2 + (q_3^p)^2}$.

3.2. Discrete element method

The discrete element method (DEM) treats particles as discrete geometric objects that interact at their points (or surfaces) of contact, permits the finite movements and rearrangements of particles and uses an explicit time-domain relaxation and integration scheme to approximate the particles' equations of motion. The method, developed by Peter Cundall, was first applied to jointed rock systems and later to granular materials [CUN 79]. Cundall had originally used the term *distinct* element method for this particular method and had used *discrete* element methods (in the plural) to designate the entire class of discontinuous particle methods (including successor methods, such as the Contact Dynamics method, the Discontinuous Deformation Analysis method, etc.). The term "discrete" element method is now generally reserved for explicit time-domain relaxation methods. The molecular dynamics (MD) methods, widely used in chemistry and fluid mechanics to model interactions of a molecular scale, are similar to the discrete element method, and the terms MD and DEM are often used interchangeably to describe the same methods. MD methods, however, do not usually include friction as an interaction mechanism.

The DEM is the most versatile of the numerical particle methods described in this chapter, as the method is suitable for rapid collisional flows and for quasi-static conditions; for any particle shape that can be expressed as a mathematical, geometric form; for systems in which deformation of the contacts and particles is an essential mechanism (e.g. dense granular materials at small strains) and for systems in which the behavior is dominated by particle rearrangements (e.g. dense materials during slow flow at large strains); for a wide variety of contact displacement–force models and even for particles that can interact without touching; for simulating both element tests and large boundary value problems; for simulating coupled problems involving fluid and heat flows within a deforming particulate medium; for accommodating the finite element (FEM) and other continuum numerical methods in coupled analyses; and for many other problems. This versatility and other attractive features have made the DEM the most widely used numerical method for simulating discrete particle systems and have led to numerous DEM codes, including commercial codes, open source codes and codes privately shared among researchers. For a recent list of codes, the reader is referred to online sites, for example [WIK 16].

As with other simulation methods, the DEM requires a means of inducing and tracking the translations and rotations of particles, detecting the proximity or contact between particles and computing the internal forces among the particles. These issues are now described, beginning with a summary of the underlying numerical algorithm. Occasional distinctions are made between slow, quasi-static simulations, in which maintaining internal force equilibrium is paramount, and dynamic simulations, in which a modest fidelity of the particle accelerations is expected. An example of the former is an element test to determine the quasi-static constitutive behavior of a representative element during slow loading (section 3.1), for which the DEM algorithm serves merely as an expedient means of maintaining local, grain-scale equilibrium while the boundaries are being moved. Simulating a bouncing ball or a twirling tennis racket is an example of the latter, which requires greater attention to the kinetics of motion.

The DEM employs a time-stepping algorithm in which information at step t is used to compute the corresponding information at step $t + 1$. At step t, a search is made for all contacts among the N particles in the assembly, and the forces and moments at these contacts are determined. The search for contacts can be a naive testing of the $N(N - 1)$ particle pairs or can make use of binning and a list of potential contacts among the near-neighbor pairs (section 3.1.5). Determining whether two particles are touching requires a geometric description of their shapes and a means of tracking the particles' orientations, as described in sections 3.1.1 and 3.1.6.

With each contact that is found, other geometric characteristics must be determined: the overlapping of the particles' surfaces at the contact, the location of the contact relative to the centers of the two particles, the contact's normal direction, the curvatures of the two surfaces and the particles' movements at the contact. The force and moment at each contact is then computed from the current overlap and from the particles' displacements or velocities during the step from $t - 1$ to t. For three-dimensional simulations, the possible relevant movements are the six objective motions identified in section 2.1: one component of displacement in the direction of the contact normal, two tangential displacements, two measures of tangential contact rolling and a single twisting of the particles about the contact normal. If the contact has a viscous character, then the relevant objective velocities must also be determined. The contact force and moment are then computed with a model of contact behavior, such as those discussed in sections 2.2–2.4. Although the normal force between two elastic particles can be fully computed from the contact overlap and the surface curvatures, computing the tangential force and contact moment requires the movement *increment* between steps $t - 1$ and t as well information on the contact's history (sections 2.3 and 2.4).

After each contact force $\mathbf{f}^c$ is computed, it is added to the other contact forces that act on the two particles, p and q, and to any external body force $\mathbf{b}^t$ at step t. The combined, net force $\overset{\star}{\mathbf{f}}^{p,t}$ on particle p at step t is:

$$\overset{\star}{f}_i^{p,t} = b_i^{p,t} + \sum_{c \in C^{p,t}} f_i^{c,p,t} \qquad [3.21]$$

where $\mathbf{f}^{c,p,t}$ is the force exerted on p by q at contact c and $C^{p,t}$ represents the full set of p's contacts at time t. The opposite contact force $\mathbf{f}^{c,q,t} = -\mathbf{f}^{c,p,t}$ is added to the forces on q. In a DEM simulation of a quasi-static process, the net force $\overset{\star}{\mathbf{f}}^{p,t}$ should, in principle, be zero (see equation [1.45]). The method relies, however, on small imbalances in the contact forces to impel particles to new positions, even during a slow quasi-static simulation. In a similar manner, the torque imbalance $\overset{\star}{\mathbf{w}}^{p,t}$ on p results from its contact forces and moments and from any external body moment:

$$\overset{\star}{w}_i^{p,t} = w_i^{p,t} + \sum_{c \in C^{p,t}} \left(e_{ijk} r_j^{c,p,t} f_k^{c,p,t} + m_i^{c,p,t} \right) \qquad [3.22]$$

where the summation for all contacts is between p and its current neighbors $c \in C^{p,t}$; $\mathbf{m}^{c,p,t}$ is the contact moment on p exerted by q at c; $\mathbf{w}^{p,t}$ is the external body moment on p; and $\mathbf{r}^{c,p,t}$ is the contact vector from the centroid of p to the contact point (Figure 1.4). Because the imbalance $\overset{\star}{\mathbf{w}}^{p,t}$ is computed in the global (inertial) frame shared by all particles, it is also written as $\overset{\star}{\mathbf{w}}^{g,p,t}$ with the "g" designating the global frame.

Once the force and moment imbalances on each particle have been computed, the particle positions and velocities are then advanced in accord with the Newton and Euler equations. Among the many choices of integration schemes [ROU 04], the standard DEM algorithm employs a leapfrog integration method (similar to the velocity Verlet method), in which the net force on particle p at step t is used to alter its mid-step velocity, which is used, in turn, to advance the position of p at step $t + 1$:

$$v_i^{p,t+1/2} = v_i^{p,t-1/2} + \frac{\Delta t}{m^p} \left(\overset{\star}{f}_i^{p,t} - \frac{1}{2} c_v^p \left(v_i^{p,t-1/2} + v_i^{p,t+1/2} \right) \right) \qquad [3.23]$$

$$x_i^{p,t+1} = x_i^{p,t} + \Delta t \, v_i^{p,t+1/2} \qquad [3.24]$$

In these equations, Δt is the integration time step, m^p is the particle mass and c_v^p is a damping constant. The choice of these values is discussed in more detail below. The damping in equation [3.23] is applied to the velocity at step t, which is taken as the average of the two mid-step velocities. This equation is rearranged as:

$$v_i^{p,t+1/2} = \frac{1}{1 + \frac{1}{2} \frac{\Delta t}{m^p} c_v^p} \left[\left(1 - \frac{1}{2} \frac{\Delta t}{m^p} c_v^p \right) v_i^{p,t-1/2} + \frac{\Delta t}{m^p} \overset{\star}{f}_i^{p,t} \right] \qquad [3.25]$$

in which the damping term $\frac{1}{2} \frac{\Delta t}{m^p} c_v^p$ is seen to slightly moderate the particle's velocity, thus providing a stabilizing influence on the algorithm (section 3.2.1).

A particle's rotation can be advanced in a similar manner, provided that the particle is either a sphere or is sufficiently symmetric so that the inertia tensor of equations [3.6]–[3.10] is isotropic (i.e. the moment of inertia is the same about all axes through the particle's center of mass, such as with spheres, cubes and tetrahedra). For such special cases:

$$\omega_i^{p,t+1/2} = \omega_i^{p,t-1/2} + \frac{\Delta t}{I^p}\left(\overset{\star}{w}_i^{p,t} - \frac{1}{2}c_\omega^p\left(\omega_i^{p,t-1/2} + \omega_i^{p,t+1/2}\right)\right) \qquad [3.26]$$

$$d\theta_i^{p,t+1} = \Delta t\, \omega_i^{p,t+1/2} \qquad [3.27]$$

In these equations, I^p is the moment of inertia of particle p, ω^p is its angular velocity vector, $d\theta^p$ is its incremental rotation and c_ω^p is the angular damping coefficient. Although equations [3.26] and [3.27] are sufficient for spheres, symmetric shapes and for most quasi-static simulations with other shapes, for full dynamic simulations in which gyroscopic effects must be considered, a more complex calculation is required for finding the rotation increment, as described below in section 3.2.3.

After updating the particles' positions and orientations at step $t + 1$ with equations [3.24] and [3.27], various housekeeping tasks are completed before looping to the next time step. These tasks include updating the list of near-neighbor particles, advancing the boundary conditions, computing the average stress (see section 1.3), etc. With spheres, the increment ω^p in equation [3.27] is sufficient for computing contact displacements and rolling in the next time step. Other shapes require the revision of particles' orientations, as expressed with rotation matrices or quaternions, and this updating of orientation is discussed in section 3.2.3 below.

3.2.1. *Time step, mass and damping*

The explicit time integration scheme used in DEM simulations imposes a critical, upper limit Δt_{crit} on the time step, and larger time steps lead to unstable and spurious results [CUN 79]. O'Sullivan and Bray [OSU 04] showed that Δt_{crit} depends on the dimension of the simulation, the number of contacts per particle, the damping constant and the arrangement of contacts, and they expressed Δt_{crit} as:

$$\Delta t_{\text{crit}} = a\,\sqrt{m/k} \qquad [3.28]$$

where m is the particle mass and k is the incremental contact stiffness (either the normal stiffness or an average of the tangential and normal stiffnesses). Factors a of 0.57 and 0.71 were determined for regular square and triangular arrays of disks. Ng [NG 06] found that reducing the time step to a value much lower than Δt_{crit} has little effect on a simulation's results. The time step in equation [3.28] depends on the particle size and is usually controlled by the smallest (least massive) particles in an assembly. Larger particles, however, will have more contacts and will be influenced by the greater combined stiffnesses of these contacts, an effect that somewhat redresses the larger time step demanded by their larger mass.

Particles with stiff contacts require a small time step, increasing the computation time. This effect is particularly acute when the intent is to capture the full dynamics of the particles' motions. However, if a simulation is intended to determine slow, quasi-static behavior and the particles' self-weights are insignificant and excluded in the simulation, then the particles' masses, although required by the DEM algorithm, need not equal the actual masses. Computation time can be greatly reduced by using an artificially large particle density in such quasi-static simulations. Such simulations can be further simplified by using a common mass for all particles, provided that the range of particle sizes is small (with a ratio of largest to smallest sizes less than, say, 3). For larger disparities in size, the larger particles can be assigned larger masses, although not in strict proportion to their volumes (since they will have more contacts).

When the contacts are modeled with nonlinear stiffnesses (as with Hertz contacts, section 2.2), the stiffness k, along with the limiting time step Δt_{crit}, will depend on the contacts' indentations and on the contact forces (see equation [2.27]). This troublesome situation can require an adaptive adjustment of the time step (or particle mass) during the course of a simulation, so that the simulation remains stable at both low and high confinement stresses.

The manner in which particles advance from one time step to the next depends on the dimensionless parameter $\Delta t / \sqrt{m/k}$, and similar simulation results are obtained when the time increment, particle mass and contact stiffness are adjusted in a manner that leaves this parameter unchanged. Because the contact stiffnesses are used in computing the bulk stress, which is often the focus of a DEM simulation, the contact properties are usually left unchanged; whereas, Δt and m can be adjusted to more convenient values that enable easier data analysis while allowing efficient computation. In the author's simulations of quasi-static behavior, the time step is routinely taken as unity, $\Delta t = 1$, and the particle mass is adjusted accordingly, so that criterion [3.28] is not violated.

The damping coefficients, c_v and c_ω, in equations [3.25]–[3.26] are usually referenced to the critical damping β: the value $\beta_v = 2\sqrt{km}$ for velocity damping c_v and the value $\beta_\omega = 2R\sqrt{kI}$ for rotational damping c_ω. As a dissipation mechanism, damping serves to reduce spurious velocity spikes and to prevent them from propagating through an assembly. Damping, although usually required, can lessen and even mask the real particle-level behavior that we intend to examine, and because it must be resisted by the boundary forces, excessive damping tends to stiffen and strengthen the bulk response [NG 06]. For this reason, minimal damping is usually recommended, and the author routinely uses a coefficient c between 0.005β and 0.03β. The relative effect of damping on the simulation results can be estimated by comparing the viscous dissipation among the N particles in an assembly with the rate of work done by the boundary forces or by the internal stress, expressed as a performance parameter I_{damping}:

$$I_{\text{damping}} = \sum_{p=1}^{N} c^p |\mathbf{v}^p|^2 \Bigg/ V(\boldsymbol{\sigma} : \dot{\boldsymbol{\varepsilon}}) \qquad [3.29]$$

where V is the assembly volume and stress-power $\boldsymbol{\sigma} : \dot{\boldsymbol{\varepsilon}} = \sigma_{ij}\dot{\varepsilon}_{ij}$ is the inner product of stress $\boldsymbol{\sigma}$ and strain rate $\dot{\boldsymbol{\varepsilon}}$ (sections 1.3 and 1.4) . Apart from the linear viscous damping embodied in c_v and c_ω, another form of damping has been proposed by Potyondy and Cundall [POT 04], in which the damping force on a particle is simply proportional to the out-of-balance force, $\overset{\star}{\mathbf{f}}{}^{p,t}$ and $\overset{\star}{\mathbf{w}}{}^{p,t}$, but acts in the opposing direction. Cundall and Strack [CUN 79] proposed a further alternative (or supplement) to the viscous damping of grains, by modifying the contact displacement–force relation to apply viscous damping directly at individual contacts. They proposed that this viscous contact damping should be operative only during elastic contact movements and should be "toggled off" during frictional slip. We must be aware that the effect of excessive damping, as measured by I_{damping}, may vary within a granular assembly and will be much greater in regions of rapid localized deformation (e.g. within shear bands and near moving boundary surfaces). Indeed, viscous resistance tends to suppress such localized deformation.

3.2.2. *Strain rate and quasi-static performance*

With quasi-static simulations, the particles should remain near equilibrium and accelerations should be minimal. Ng [NG 06] proposed a dimensionless measure of force imbalance relative to the average contact force:

$$I_{\mathrm{uf}} = \sqrt{\frac{1}{N}\sum_{p=1}^{N}|\overset{\star}{\mathbf{f}}^{p}|^{2}\bigg/\frac{1}{M}\sum_{c=1}^{M}|\mathbf{f}^{c}|^{2}} \qquad [3.30]$$

where N and M are the numbers of particles and contacts, $\overset{\star}{\mathbf{f}}^{p}$ is the unbalanced force on particle p (equation [3.21]) and $\mathbf{f}^{c}$ is the force at contact c. A similar parameter can be computed for the average of the imbalances of moments $\overset{\star}{\mathbf{w}}^{p}$, such that the denominator in equation [3.30] is compiled from the magnitudes of the moments of force, $|\mathbf{r}^{c} \times \mathbf{f}^{c} + \mathbf{m}^{c}|$. The performance parameter I_{uf} should be small (certainly less than a few percent) when simulating quasi-static conditions. For contacts with a linear contact stiffness k, Suzuki and Kuhn [SUZ 14] showed that I_{uf} is proportional to the strain increment $\Delta\varepsilon$ that is advanced with each time step ($\Delta\varepsilon = \dot{\varepsilon}\Delta t$) and is inversely proportional to the average ratio of contact indentation ζ and particle radius R:

$$I_{\mathrm{uf}} \propto \frac{\Delta\varepsilon}{\zeta/R} \quad \text{and} \quad I_{\mathrm{uf}} \propto \frac{\Delta\varepsilon}{pR/k} \qquad [3.31]$$

and is inversely proportional to the ratio of the mean stress p and the contact stiffness k (the product pR is replaced with p for two-dimensional simulations). In other words, slow deformation is imperative when simulating quasi-static conditions.

Da Cruz *et al.* [DAC 05] proposed another parameter for measuring attainment of quasi-static conditions. Their dimensionless inertial number is:

$$I_{\mathrm{inertia}} = \dot{\varepsilon}\,\sqrt{m/(pD)} \qquad [3.32]$$

where m is the average particle mass, and p and D are the mean stress and average particle diameter. The inertial number gives the relative dominance of inertial and contact forces (for two-dimensional simulations, the denominator pD is replaced with p). Small values ($I < 0.01$) correspond to the quasi-static regime of behavior in which the evolving stress is sensitive to particle stiffness and to contact friction. Values larger than 0.2 are characteristic of the collisional domain in which the particles' inertias predominantly influence the internal stress and bulk viscosity. The author routinely maintains values of I_{inertia} less than 0.0001 for quasi-static simulations. A parameter closely related to the inertial number is simply the ratio of the particles' kinetic energy (the sum of $\frac{1}{2}m^p|\mathbf{v}^p|^2$ values, taken from equation [3.25]) and the combined elastic energy among the particle contacts.

3.2.3. *Kinetics of non-symmetric particle shapes*

Equations [3.26] and [3.27] suffice for spheres and for other symmetric shapes that have a common moment of inertia about all centroid axes. The equations are also adequate for quasi-static simulations in which the particles' rotational inertias are small. The situation is more complex for the rapid flow of particles that have non-symmetric shapes and in which the dynamic tumbling and twirling of the particles must be faithfully tracked. The torque imbalance on a particle, the vector $\overset{\star}{\mathbf{w}}{}^p$ in equation [3.22], changes the particle's angular momentum and impels its rotation. The angular momentum vector $\mathbf{L}^p$ of particle p is the product of its inertia tensor $\mathbf{I}^p$ and the angular velocity vector ω^p, which will have different components in the global ("g", inertial) and local ("l", body) frames:

$$\mathbf{L}^{g,p} = \mathbf{I}^{g,p} \cdot \omega^{g,p} \quad \text{and} \quad \mathbf{L}^{l,p} = \mathbf{I}^{l,p} \cdot \omega^{l,p} \qquad [3.33]$$

The two inertia tensors, $\mathbf{I}^{g,p}$ and $\mathbf{I}^{l,p}$, are defined in equations [3.6]–[3.9] (the momenta $\mathbf{L}^{g,p}$ and $\mathbf{L}^{l,p}$ should not be confused with the velocity gradient $\mathbf{L}$ in equation [1.93]). The angular velocity in the global and local frames is related through the rotation matrix $\mathbf{R}^p$:

$$\omega^{g,p} = \mathbf{R}^p \cdot \omega^{l,p} \quad \text{and} \quad \omega^{l,p} = (\mathbf{R}^p)^{-1} \cdot \omega^{g,p} \qquad [3.34]$$

where $\mathbf{R}^p$ is defined in equations [3.4] and [3.18]. With these quantities, Newton's equation of three dimensional rigid rotation is:

$$\overset{\star}{\mathbf{w}}{}^{g,p} = \frac{d\mathbf{L}^{g,p}}{dt} = \frac{d}{dt}(\mathbf{I}^{g,p} \cdot \omega^{g,p}) \qquad [3.35]$$

The torque imbalance $\overset{\star}{\mathbf{w}}{}^{g,p}$ is expressed in the global frame and is the same imbalance as that in equation [3.22]. This expression replaces equations [3.26]–[3.27] for particles with non-symmetric shapes. Solving equation [3.35] to update the angular velocity is impeded by the fact that both the angular velocity and the global inertia tensor are often simultaneously changing. An approximate solution is commonly applied in computer animations and games, by assuming that inertia tensor $\mathbf{I}^{g,p}$ is constant during a time step Δt and is only updated at the end of the time step [HEC 97, BAR 01]. The sequence of computational steps for approximating and updating the rotational velocity, the rotation matrix and the inertia tensor is as follows:

$$\omega^{g,p,t+1/2} = \left(\mathbf{I}^{g,p,t}\right)^{-1}\left[\mathbf{I}^{g,p,t}\cdot\omega^{g,p,t-1/2} + \Delta t\,\overset{\star}{\mathbf{w}}^{g,p,t}\right] \qquad [3.36]$$

$$\omega^{g,p,t} = \tfrac{1}{2}\left(\omega^{g,p,t-1/2} + \omega^{g,p,t+1/2}\right) \qquad [3.37]$$

$$\mathbf{R}^{p,t+1} = \mathbf{R}^{p,t} + \Delta t\left(\omega^{g,p,t} * \mathbf{R}^{p,t}\right) \qquad [3.38]$$

$$\mathbf{I}^{g,p,t+1} = \mathbf{R}^{p,t+1}\cdot\mathbf{I}^{1,p}\cdot(\mathbf{R}^{p,t+1})^{-1} \qquad [3.39]$$

where equations [3.11] and [3.9] have been applied to advance the particle's rotation matrix and to shift from its stationary local inertia tensor $\mathbf{I}^{1,p}$ to the revised global inertia $\mathbf{I}^{g,p}$. Between equations [3.38] and [3.39], the rotation matrix $R^{p,t+1}$ must be re-orthogonalized, so that it can be used in the next time step for contact detection and calculation of contact forces (see equations [3.5]). Again, this algorithm is approximate, as it fails to account for the concurrent changes in ω^g and $\mathbf{I}^g$.

This shortcoming is avoided by expressing the kinetics of equation [3.35] in the body (rotating) frame rather than in the stationary inertial frame. The resulting Euler's equation is:

$$\overset{\star}{\mathbf{w}}^{1,p} = \frac{d\mathbf{L}^{1,p}}{dt} + \omega^{1,p}\times\mathbf{L}^{1,p} = \mathbf{I}^{1,p}\cdot\frac{d\omega^{1,p}}{dt} + \omega^{1,p}\times\left(\mathbf{I}^{1,p}\cdot\omega^{1,p}\right) \qquad [3.40]$$

where $\mathbf{L}^{1,p}$ is defined in equation [3.33$_2$] and $\overset{\star}{\mathbf{w}}^{1,p}$ is the torque imbalance of equation [3.22], but referenced to the body frame (see equation [3.34$_2$]). Euler's equation has the advantage of using a local inertia tensor $\mathbf{I}^{1,p}$ that does not change, even as a particle rotates within the global frame. The difficulty lies in solving the rotation increment $d\omega^{1,p}$. When viewed as a differential equation for the three components of rotation, $\omega_i^{1,p}$, equation [3.40] is nonlinear and coupled. By aligning the body frame with the directions of the particle's principal inertias, the equation is separated into the three components of $d\omega^{1,p}$ as follows:

$$d\omega_1^{1,p}/dt = \left[\overset{\star}{w}_1^{1,p} + \omega_2^{1,p}\omega_3^{1,p}\left(I_2^{1,p} - I_3^{1,p}\right)\right]\Big/I_1^{1,p} \qquad [3.41]$$

$$d\omega_2^{1,p}/dt = \left[\overset{\star}{w}_2^{1,p} + \omega_1^{1,p}\omega_3^{1,p}\left(I_3^{1,p} - I_1^{1,p}\right)\right]\Big/I_2^{1,p} \qquad [3.42]$$

$$d\omega_3^{1,p}/dt = \left[\overset{\star}{w}_3^{1,p} + \omega_1^{1,p}\omega_2^{1,p}\left(I_1^{1,p} - I_2^{1,p}\right)\right]\Big/I_3^{1,p} \qquad [3.43]$$

where $I_i^{1,p}$ are the principal inertias (see equation [3.10]). Numerically solving these equations requires a predictor–corrector approach, in which initial values of $\omega^{1,p}$ are used for an initial prediction of $d\omega^{1,p}$, and then this prediction is used for a temporary updating of $\omega^{1,p}$ which is then used for the final calculation of $d\omega^{1,p}$. Several such predictor–corrector algorithms have been suggested in the literature [WAL 93, ROZ 10, LIM 14]. The updated angular velocity is then used in equations [3.18]–[3.20] to update the orientation quaternion $\overset{\circ}{\mathbf{q}}^p$ and the corresponding orientation matrix $\mathbf{R}^p$.

3.3. Stiffness matrix methods

The discrete element method (DEM) approximates the solution of Newton's equations for an entire granular system by considering the local equilibrium of each particle and iteratively adjusting its position until it and all other particles approach near-equilibrium (section 3.2). This scheme is similar to using iterative Gauss–Seidel relaxation to approximate the solution of a large system of linear equations, although the problem is far more difficult with granular assemblies, since equilibrium is sought within a continually changing system, and the contact interactions are usually nonlinear.

Other methods attack the problem more directly by creating and solving the full system of equations for the contact forces and particle movements. These methods linearize the equations and express them in matrix form. The discontinuous deformation analysis (DDA) method, originally intended for fractured rock-block systems, is one such approach. The original method, developed by Shi [SHI 88, SHI 93], assumed that the contacts between frictional blocks were rigid and non-penetrating but that the blocks were deformable. By treating the blocks as independent but deforming bodies, DDA tracks the dynamics of an entire block system by applying Newton's equation of motion in a matrix form. Ke and Bray [KE 95] extended the DDA method to dense systems of rigid disks having deformable contacts. Rather than tracking the full dynamics of a granular assembly, other matrix methods are primarily concerned with quasi-static incremental movements and addressing questions of anisotropy, incremental plasticity, stability and uniqueness. These methods include the matrix stiffness method of Kishino [KIS 88, KIS 03], which he called the granular element method (GEM), and the methods of Bagi [BAG 07] and Kuhn and Chang [KUH 06].

The incrementally nonlinear nature of contact stiffness creates a profound difficulty with these matrix-based methods. With simple linear-frictional contacts (section 2.3), the incremental stiffness of a contact that has reached the frictional limit (i.e. $|\mathbf{f}^t| = \mu f^n$) has two branches: one elastic and one with sliding. A Cattaneo–Mindlin contact is even more complex, as it exhibits a smooth transition toward sliding and is incrementally nonlinear even before the frictional limit is reached (section 2.3.1). Stiffness in the normal direction can also be incrementally nonlinear: for example, with a non-Hertzian contact that is on the verge of touching (i.e. grazing contact), the normal stiffness can abruptly transition from zero to a finite stiffness. When a matrix approach is used, the difficulties associated with friction and with grazing require an iterative process to determine which contacts are subject to such incremental nonlinearity of stiffness, for example determining which contacts are either sliding or are in a state of incipient sliding. After identifying these contacts, we must then determine the appropriate stiffness to apply with each of these contacts during the next increment of movement. This section describes the incremental stiffness matrix and the manner in which it must be adjusted when such incremental nonlinearity is present.

The incremental stiffness matrix can be developed by considering the quasi-static equilibrium of an entire assembly or of a smaller cluster of particles during an increment of loading. The particle positions, contact forces and loading history are assumed known at the current time t. If some of the particle motions or the external particle forces are prescribed for the increment dt (for example, boundary movements and forces), the problem is to find the remaining, unknown increments of motion and force. The particle motions are assumed to be governed by the mechanics of rigid bodies with compliant contacts: particle motions produce contact deformations (the kinematic problem); contact deformations produce increments in the contact forces (the contact constitutive problem); the forces on each particle must comply with Newton's equations (the equilibrium problem). The incremental stiffness equation for a three-dimensional system of N particles undergoing a quasi-static near-equilibrium process is:

$$[\mathbf{H}]_{6N \times 6N} \begin{bmatrix} d\mathbf{u} \\ \hdots \\ d\theta \end{bmatrix}_{6N \times 1} = \begin{bmatrix} d\mathbf{b} \\ \hdots \\ d\mathbf{w} \end{bmatrix}_{6N \times 1} \qquad\qquad [3.44]$$

where $[\mathbf{H}]$ is the global incremental stiffness matrix; vector $[d\mathbf{u}/d\theta]$ is the $6N \times 1$ list of the three incremental displacements and the three incremental rotations for each of the N particles (Figure 2.1); vector $[d\mathbf{b}/d\mathbf{w}]$ is a list of the six infinitesimal increments of external force and moment applied to the reference points χ of each of the N particles (Figure 3.4). The ordering of the individual displacements, rotations, forces and moments in the vectors $[d\mathbf{u}/d\theta]$ and $[d\mathbf{b}/d\mathbf{w}]$ is arbitrary, although items within the vector on the left should correspond with their complementary quantities on the right. Although the displacements and rotations on the left side of equation [3.44] may appear as unknown, when the boundary conditions prescribe certain displacements and rotations, they and their complementary unknown forces and moments must switch sides to permit solution of all unknown quantities. The equations are then solved with standard methods of computational linear algebra, as with the finite element method.

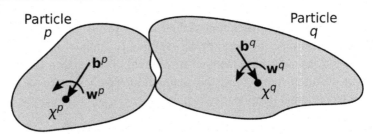

Figure 3.4. *External forces and moments on particles*

The stiffness matrix $[\mathbf{H}]$ can be assembled from the stiffness matrices of the assembly's elemental units, which are the individual contacts between particle pairs. Consider two representative particles, p and q, that are in contact.

The incremental stiffness contributed by this one contact can be expressed in matrix form as:

$$
\begin{bmatrix} \mathbf{H}^{p-p} & \mathbf{H}^{p-q} \\ \hdashline \mathbf{H}^{q-p} & \mathbf{H}^{q-q} \end{bmatrix}_{12 \times 12}
\begin{bmatrix} du^p \\ d\theta^p \\ \hdashline du^q \\ d\theta^q \end{bmatrix}_{12 \times 1}
=
\begin{bmatrix} db^{p,\,pq} \\ dw^{p,\,pq} \\ \hdashline db^{q,\,qp} \\ dw^{q,\,qp} \end{bmatrix}_{12 \times 1}
\qquad [3.45]
$$

where du^p, du^q, $d\theta^p$ and $d\theta^q$ are the translations and rotations of p and q (Figure 2.1). Equation [3.45] expresses the effect that the single contact will have on the equilibrium of the two particles. With non-convex particles, multiple contacts are possible between a particle pair. In this case, the 12×12 matrix in equation [3.45] is a sum of the 12×12 matrices associated with the multiple contacts c between p and q (Figure 1.6(c)).

The external force increments on the right of equation [3.45] must be combined (assembled) with the forces that are produced by the other contacts in an assembly or cluster. The stiffness matrices of all M contacts within the cluster can be assembled into the global stiffness matrix $[\mathbf{H}]$ of equation [3.44]. The following "$\leadsto$" notation will be used to indicate the assembly of the full matrix $[\mathbf{H}]$ from its contributing parts:

$$
\begin{bmatrix} \mathbf{H}^{p-p} & \mathbf{H}^{p-q} \\ \hdashline \mathbf{H}^{q-p} & \mathbf{H}^{q-q} \end{bmatrix}_{12 \times 12}
\leadsto [\mathbf{H}]_{6N \times 6N}
\qquad [3.46]
$$

$$
\begin{bmatrix} du^p \\ d\theta^p \\ \hdashline du^q \\ d\theta^q \end{bmatrix}_{12 \times 1}
\leadsto \begin{bmatrix} du \\ d\theta \end{bmatrix}_{6N \times 1}
\quad \text{and} \quad
\begin{bmatrix} db^{p,\,pq} \\ dw^{p,\,pq} \\ \hdashline db^{q,\,qp} \\ dw^{q,\,qp} \end{bmatrix}_{12 \times 1}
\leadsto \begin{bmatrix} db \\ dw \end{bmatrix}_{6N \times 1}
\qquad [3.47]
$$

meaning that the elements of sub-matrix $\mathbf{H}^{x-y}$ are added to the corresponding rows and columns of the $6N \times 6N$ matrix $[\mathbf{H}]$: the rows of $[\mathbf{H}]$ that correspond to the x forces in the global vector $[db/dw]$ and the columns of $[\mathbf{H}]$ that correspond to the y displacements in the global vector $[du/d\theta]$.

The 12×12 stiffness matrix in equation [3.45] is derived by considering all contact forces $\mathbf{f}$ and all contact moments $\mathbf{m}$ that are exerted on the single particle p by its neighbors (Figures 1.6, 1.7 and 3.4). These forces are assumed to be in equilibrium at time t:

$$-\sum_{q\in Q^p} \mathbf{f}^{pq} = \mathbf{b}^p , \quad -\sum_{q\in Q^p} (\mathbf{r}^{pq} \times \mathbf{f}^{pq} + \mathbf{m}^{pq}) = \mathbf{w}^p \qquad [3.48]$$

where $\mathbf{r}^{pq}$ connects the center of p to its contact with q (Figure 1.4) and the summations include all of p's neighbors, $q \in Q^p$ (Figure 1.6(a)). For a non-convex particle that has multiple contacts with a neighbor q, we can alternatively use summations of all of p's contacts, $c \in C^p$, as in Figure 1.6(b).

Because equation [3.44] is an incremental expression, equilibrium must be cast in an incremental form:

$$-\sum_{q\in Q^p} d\mathbf{f}^{pq} = d\mathbf{b}^p , \quad -\sum_{q\in Q^p} (d\mathbf{r}^{pq} \times \mathbf{f}^{pq} + \mathbf{r}^{pq} \times d\mathbf{f}^{pq} + d\mathbf{m}^{pq}) = d\mathbf{w}^p [3.49]$$

which include increments in the contact (internal) forces, $d\mathbf{f}^{pq}$ and $d\mathbf{m}^{pq}$, increments in the external forces, $d\mathbf{b}^p$ and $d\mathbf{w}^p$ and incremental changes in the contact positions, $d\mathbf{r}^{pq}$. Equation [3.49] is the incremental counterpart of equations [1.59] and [1.60], but now includes external forces. Equilibrium is seen to be altered by changes in the contact location (i.e. the $d\mathbf{r}^{pq}$ increment) as well as by changes of the contact and external forces. The change $d\mathbf{r}^{pq}$ is a geometric effect that makes the relationship between movements and forces path dependent (non-conservative), apart from any path dependence that might result from an inelastic contact mechanism.

By applying equation [3.49], the author has shown [KUH 06] that the $6N \times 6N$ global stiffness matrix $[\mathbf{H}]$ in equation [3.44] is the sum of two parts: one part that is geometric and originates in the current contact forces and in the incremental geometric changes of the particle arrangement, and another part that is mechanical and is due to incremental deformations at the contacts:

$$[\mathbf{H}] = [\mathbf{H}^g] + [\mathbf{H}^m] \qquad [3.50]$$

in which all three matrices have size $6N \times 6N$. Although the mechanical stiffness $[\mathbf{H}^m]$ is an approximation of $[\mathbf{H}]$ and may be sufficient for most numerical simulations, analyzes of internal instability and of displacement bifurcations require inclusion of the geometric effects embodied in $[\mathbf{H}^g]$. These geometric effects have three matrix contributions:

$$[\mathbf{H}^g] = [\mathbf{H}^{g-1}] + [\mathbf{H}^{g-2}] + [\mathbf{H}^{g-3}]$$ [3.51]

which are described below.

The issue of objectivity, discussed in section 2.1, must also be addressed in regard to the incremental form of equilibrium in equation [3.49]. Contact deformations, which produce *mechanical* changes in the contact forces, must be computed from the relative, objective motions of the two particles at their points of contact: the contact deformation, rolling, and twisting discussed in section 2.1. The issue of objectivity can be resolved by temporarily replacing the non-objective incremental quantities in equation [3.49] with the following "δ" objective increments. These increments would be viewed by an observer attached to (and rotating with) particle p (or alternatively, with particle q):

$$d\mathbf{r}^{pq} = \delta\mathbf{r}^{pq} + d\boldsymbol{\theta}^p \times \mathbf{r}^{pq} \qquad d\mathbf{r}^{qp} = \delta\mathbf{r}^{qp} + d\boldsymbol{\theta}^q \times \mathbf{r}^{qp}$$ [3.52]

$$d\mathbf{f}^{pq} = \delta\mathbf{f}^{pq} + d\boldsymbol{\theta}^p \times \mathbf{f}^{pq} \qquad d\mathbf{f}^{qp} = \delta\mathbf{f}^{qp} + d\boldsymbol{\theta}^q \times \mathbf{f}^{qp}$$ [3.53]

$$d\mathbf{m}^{pq} = \delta\mathbf{m}^{pq} + d\boldsymbol{\theta}^p \times \mathbf{m}^{pq} \qquad d\mathbf{m}^{qp} = \delta\mathbf{m}^{qp} + d\boldsymbol{\theta}^q \times \mathbf{m}^{qp}$$ [3.54]

$$d\mathbf{b}^p = \delta\mathbf{b}^p + d\boldsymbol{\theta}^p \times \mathbf{b}^p \qquad d\mathbf{b}^q = \delta\mathbf{b}^q + d\boldsymbol{\theta}^q \times \mathbf{b}^q$$ [3.55]

$$d\mathbf{w}^p = \delta\mathbf{w}^p + d\boldsymbol{\theta}^p \times \mathbf{w}^p \qquad d\mathbf{w}^q = \delta\mathbf{w}^q + d\boldsymbol{\theta}^q \times \mathbf{w}^q$$ [3.56]

Because the "δ" increments are viewed from their respective particles, they will be different for p and q. For example, although $d\mathbf{f}^{pq} = -d\mathbf{f}^{qp}$, the increment $\delta\mathbf{f}^{pq}$ is not necessarily equal to $-\delta\mathbf{f}^{qp}$. Significantly, the incremental equilibrium of equation [3.49] applies to both the "δ" and "d" quantities [KUH 06]:

$$-\sum_{q \in Q^p} \delta\mathbf{f}^{pq} = \delta\mathbf{b}^p, \qquad -\sum_{q \in Q^p} (\delta\mathbf{r}^{pq} \times \mathbf{f}^{pq} + \mathbf{r}^{pq} \times \delta\mathbf{f}^{pq} + \delta\mathbf{m}^{pq}) = \delta\mathbf{w}^p$$ [3.57]

The first geometric contribution $[\mathbf{H}^{g-1}]$ arises from the $\delta\mathbf{r}^{pq} \times \mathbf{f}^{pq}$ term in equation [3.57]. The increment of the contact vector, $\delta\mathbf{r}^{pq}$, has components normal to and tangent to the contact surfaces. The normal component is half the increment $d\mathbf{u}^{c,\text{rigid},n}$ given in equation [2.3₁] and is the negative of the half-increment of indentation ζ of equation [2.24]. The tangential component is the term $\mathbf{t}^{c,p} ds^p$ in equation [2.10], which depends on the curvatures of the two contacting surfaces:

$$\delta \mathbf{r}^{pq} = \frac{1}{2} d\mathbf{u}^{c,\text{rigid},n} + \mathbf{t}^{c,p} ds^p \qquad [3.58]$$

This increment represents a change in the contact position, as viewed by an observer attached to and rotating with p as the contact crawls across the surface of p. Because indentations are small relative to the particle size, the effect of the normal component is usually very small. The tangential contribution, however, can be quite large due to the vigorous rolling and sliding that often occur between particles.

The terms $\delta \mathbf{r}^{pq} \times \mathbf{f}^{pq}$ and $\delta \mathbf{r}^{qp} \times \mathbf{f}^{qp}$ that affect the equilibrium of particles p and q are assembled into the first geometric stiffness:

$$\begin{bmatrix} \delta \mathbf{r}^{pq} \times \mathbf{f}^{pq} \\ \hdashline \delta \mathbf{r}^{qp} \times \mathbf{f}^{qp} \end{bmatrix}_{12 \times 12} \leadsto \left[\mathbf{H}^{\mathrm{g}-1} \right]_{6N \times 6N} \qquad [3.59]$$

such that terms on the left are added into the proper rows and columns of the $6N \times 6N$ matrix $[\mathbf{H}^{\mathrm{g}-1}]$. Note that the 12×12 matrix on the left when multiplied by the twelve movements $\{d\mathbf{u}^p, d\mathbf{u}^q, d\theta^p, d\theta^q\}$ yields increments of force and moment on p and q, as per equation [3.57]; equations [2.3$_1$] and [2.10]; and the similar equations for q. In this manner, the six equilibrium equations [3.57] of each particle can be gathered into the $6N$ equations of all particles, by collecting the contact force increments, $\delta \mathbf{f}^{pq}$, $\delta \mathbf{f}^{qp}$, $\delta \mathbf{m}^{pq}$ and $\delta \mathbf{m}^{qp}$, from each of the M contacts:

$$\left[\mathbf{H}^{\mathrm{g}-1} \right]_{6N \times 6N} \begin{bmatrix} d\mathbf{u} \\ \hdashline d\theta \end{bmatrix}_{6N \times 1} - \left[\mathbf{A}_1 \right]_{6N \times 2(6M)} \begin{bmatrix} \delta \mathbf{f} \\ \hdashline \delta \mathbf{m} \end{bmatrix}_{2(6M) \times 1} = \begin{bmatrix} \delta \mathbf{b} \\ \hdashline \delta \mathbf{w} \end{bmatrix}_{6N \times 1} \qquad [3.60]$$

When assembling the contact forces and moments in this equation, the contact forces $\delta \mathbf{f}^{pq}$ and $\delta \mathbf{f}^{qp}$ are treated as distinct objects, since $\delta \mathbf{f}^{pq}$ and $\delta \mathbf{m}^{pq}$ are not usually equal to their qp counterparts, $-\delta \mathbf{f}^{qp}$ and $-\delta \mathbf{m}^{qp}$. This distinction leads to a total of $2(6M)$ contact force/moment components among the M contacts (three components of both $\delta \mathbf{f}$ and $\delta \mathbf{m}$ for each of the "pq" and

"qp" varieties). The statics matrix $[\mathbf{A}_1]$ combines these contact forces and moments, as with the $\delta\mathbf{f}^{pq}$ and $\delta\mathbf{m}^{pq}$ sums of equations [3.57]:

$$
-\begin{bmatrix}
\delta\mathbf{f}^{pq} \\
\hline
\mathbf{r}^{pq} \times \delta\mathbf{f}^{pq} + \delta\mathbf{m}^{pq} \\
\hline
\delta\mathbf{f}^{qp} \\
\hline
\mathbf{r}^{qp} \times \delta\mathbf{f}^{qp} + \delta\mathbf{m}^{qp}
\end{bmatrix}_{12\times 1}
\rightsquigarrow
-[\,\mathbf{A}_1\,]_{6N\times 2(6M)}
\begin{bmatrix}
\delta\mathbf{f}^{pq} \\
\delta\mathbf{m}^{pq} \\
\hline
\delta\mathbf{f}^{qp} \\
\delta\mathbf{m}^{qp}
\end{bmatrix}_{2(6M)\times 1}
\qquad [3.61]
$$

A second geometric effect is due to changes in the *directions* of the contact forces as their contact points shift across the surfaces of the particles. To achieve the incremental form of equation [3.44], the product $[\mathbf{A}_1][\delta\mathbf{f}/\delta\mathbf{m}]$ in equations [3.60] and [3.61] must be expressed in terms of the $6N$ particle movements $[d\mathbf{u}/d\boldsymbol{\theta}]$. The increments of a single contact's force and moment will depend, of course, on the deformation of the two particles at their contact, as discussed in sections 2.2–2.4, but the increments will also depend on any change in the orientation of the pair's contact plane. The increments in contact force and moment that are caused exclusively by deformation are designated as $\eth\mathbf{f}^{pq}$ and $\eth\mathbf{m}^{pq}$. As viewed by a global (inertial) observer, the combined increments of force and moment at the contact pq are:

$$
d\mathbf{f}^{pq} = \eth\mathbf{f}^{pq} + \mathbf{f}^{pq} \times (d\mathbf{n}^{pq} \times \mathbf{n}^{pq}) - \tfrac{1}{2}\left[(d\boldsymbol{\theta}^p + d\boldsymbol{\theta}^q) \cdot \mathbf{n}^{pq}\right]\mathbf{f}^{pq} \times \mathbf{n}^{pq}
$$
$$[3.62]$$

$$
d\mathbf{m}^{pq} = \eth\mathbf{m}^{pq} + \mathbf{m}^{pq} \times (d\mathbf{n}^{pq} \times \mathbf{n}^{pq}) - \tfrac{1}{2}\left[(d\boldsymbol{\theta}^p + d\boldsymbol{\theta}^q) \cdot \mathbf{n}^{pq}\right]\mathbf{m}^{pq} \times \mathbf{n}^{pq}
$$
$$[3.63]$$

The geometric effects $\mathbf{f}^{pq} \times (d\mathbf{n}^{pq} \times \mathbf{n}^{pq})$ and $\mathbf{m}^{pq} \times (d\mathbf{n}^{pq} \times \mathbf{n}^{pq})$ are the force increments produced by a rotation (tilting) of the contact plane, as seen by an inertial "d" observer. These terms are described in section 2.1 (page 72 and equations [2.7] and [2.12]). The final, subtracted terms in equations [3.62] and [3.63] are geometric effects produced by a rigid-body twirling of the particle pair, as described near equation [2.20].

Equations [3.62] and [3.63] can also be written in terms of the co-rotated, objective "δ" vectors, as required in equations [3.57] and [3.60] [KUH 06]:

$$
\delta\mathbf{f}^{pq} = \eth\mathbf{f}^{pq} + \mathbf{f}^{pq} \times (\delta\mathbf{n}^{pq} \times \mathbf{n}^{pq}) - \tfrac{1}{2}\left((d\boldsymbol{\theta}^q - d\boldsymbol{\theta}^p) \cdot \mathbf{n}^{pq}\right)\mathbf{f}^{pq} \times \mathbf{n}^{pq}
$$
$$[3.64]$$

$$
\delta\mathbf{m}^{pq} = \eth\mathbf{m}^{pq} + \mathbf{m}^{pq} \times (\delta\mathbf{n}^{pq} \times \mathbf{n}^{pq}) - \tfrac{1}{2}\left((d\boldsymbol{\theta}^q - d\boldsymbol{\theta}^p) \cdot \mathbf{n}^{pq}\right)\mathbf{m}^{pq} \times \mathbf{n}^{pq}
$$
$$[3.65]$$

Complementary equations for $\delta\mathbf{f}^{qp}$ and $\delta\mathbf{m}^{qp}$ are similar, but with the "p" and "q" order exchanged. The relationship between increments $d\mathbf{n}^{pq}$ and $\delta\mathbf{n}^{pq}$ is:

$$dn^{pq} = \delta n^{pq} + d\theta^{p} \times n^{pq} \quad \text{and} \quad dn^{qp} = \delta n^{qp} + d\theta^{q} \times n^{qp} \qquad [3.66]$$

(see equations [3.52]–[3.56]), and comparing these equations with equations [2.7] and [2.8] shows that:

$$\delta n^{pq} = -\mathbf{K}^{p} \cdot (\mathbf{t}^{c,p} ds^{p}) \quad \text{and} \quad \delta n^{qp} = -\mathbf{K}^{q} \cdot (\mathbf{t}^{c,q} ds^{q}) \qquad [3.67]$$

where $\mathbf{t}^{c,p} ds^{p}$ and $\mathbf{t}^{c,q} ds^{q}$ are defined in equations [2.10] and [2.11]. With these expressions, the last two terms in equations [3.64] and [3.65] are geometric effects that can be collected into a matrix that encompasses all M contacts:

$$\begin{bmatrix}
\mathbf{f}^{pq} \times (\delta\mathbf{n}^{pq} \times \mathbf{n}^{pq}) - \frac{1}{2}\left(\delta\theta^{pq,\,\text{rigid}} \cdot \mathbf{n}^{pq}\right)\mathbf{f}^{pq} \times \mathbf{n}^{pq} \\
\hline
\mathbf{m}^{pq} \times (\delta\mathbf{n}^{pq} \times \mathbf{n}^{pq}) - \frac{1}{2}\left(\delta\theta^{pq,\,\text{rigid}} \cdot \mathbf{n}^{pq}\right)\mathbf{m}^{pq} \times \mathbf{n}^{pq} \\
\hline
\mathbf{f}^{qp} \times (\delta\mathbf{n}^{qp} \times \mathbf{n}^{qp}) - \frac{1}{2}\left(\delta\theta^{qp,\,\text{rigid}} \cdot \mathbf{n}^{qp}\right)\mathbf{f}^{qp} \times \mathbf{n}^{qp} \\
\hline
\mathbf{m}^{qp} \times (\delta\mathbf{n}^{qp} \times \mathbf{n}^{qp}) - \frac{1}{2}\left(\delta\theta^{qp,\,\text{rigid}} \cdot \mathbf{n}^{qp}\right)\mathbf{m}^{qp} \times \mathbf{n}^{qp}
\end{bmatrix}_{12\times 1}$$

$$\rightsquigarrow [\mathbf{A}_2]_{2(6M)\times 6N} \begin{bmatrix} d\mathbf{u} \\ \hline d\theta \end{bmatrix}_{6N\times 1} \qquad [3.68]$$

where $\delta\theta^{pq,\,\text{rigid}}$ and $\delta\theta^{qp,\,\text{rigid}}$ are the relative rotations given in equation [2.18], noting that $\delta\theta^{pq,\,\text{rigid}} = -\delta\theta^{qp,\,\text{rigid}}$.

The contact force and moment increments, $\delta\mathbf{f}^{pq}$ and $\delta\mathbf{m}^{pq}$, in equations [3.64] and [3.65] are the objective *mechanical* increments caused by deformations of the particles, and they depend on the six objective contact movements that were discussed in section 2.1. These objective movements include the relative movement $d\mathbf{u}^{pc,\text{rigid}}$ of materials points in the two particles near their contact (equation [2.2]), the rolling $d\mathbf{u}^{pq,\text{roll}}$ of the particles (equation [2.13]) and the twisting rotation $d\theta^{pq,\text{twist}}$ of the particles about the contact normal (equation [2.17]). In place of the rolling $d\mathbf{u}^{pq,\text{roll}}$, we can use the relative rotation of the two particles, $d\theta^{pq,\,\text{rigid}}$, of equation [2.18].

In general, the incremental contact stiffness is a functional relation of the form:

$$\eth\mathbf{f}^{pq} = \mathbf{F}^{pq}\left(d\mathbf{u}^{pq,\text{rigid}}, d\theta^{pq,\,\text{rigid}}\right) \tag{3.69}$$

$$\eth\mathbf{m}^{pq} = \mathbf{M}^{pq}\left(d\mathbf{u}^{pq,\text{rigid}}, d\theta^{pq,\,\text{rigid}}\right) \tag{3.70}$$

which includes the contact models discussed in sections 2.2–2.4. Note that the relative rotation of equation [2.18] is used in place of the rolling and twisting movements of equations [2.13], [2.16] and [2.17], although these latter motions could also be used in equations [3.69]–[3.70]. The contact stiffness can be incrementally nonlinear, with branches of slip, micro-slip and elastic behaviors that depend on the direction of loading. The stiffness can be history dependent, such that the stiffness of a contact depends not only on the current contact force but also on the history of movements leading to this force. With this understanding, the general stiffness relations in equations [3.69] and [3.70] are collected for all M contacts into the matrix form:

$$\begin{bmatrix} \eth\mathbf{f}^{pq} \\ \hline \eth\mathbf{m}^{pq} \\ \hline \eth\mathbf{f}^{qp} \\ \hline \eth\mathbf{m}^{qp} \end{bmatrix}_{2(6M)\times 1} = \begin{bmatrix} \mathbf{F}^{pq} \\ \hline \mathbf{M}^{pq} \\ \hline \mathbf{F}^{qp} \\ \hline \mathbf{M}^{qp} \end{bmatrix}_{2(6M)\times 6M} \begin{bmatrix} d\mathbf{u}^{pq,\text{rigid}} \\ d\theta^{pq,\,\text{rigid}} \end{bmatrix}_{6M\times 1} \tag{3.71}$$

noting that $\eth\mathbf{f}^{pq} = -\eth\mathbf{f}^{qp}$, $\eth\mathbf{m}^{pq} = -\eth\mathbf{m}^{qp}$, $\mathbf{F}^{pq} = -\mathbf{F}^{qp}$ and $\mathbf{M}^{pq} = -\mathbf{M}^{qp}$. The stiffness coefficients, $\mathbf{F}$ and $\mathbf{M}$, can depend on the movement *directions* of $d\mathbf{u}^{\text{rigid}}$ and $d\theta^{\text{rigid}}$, in accord with their incrementally nonlinear character (e.g. equation [2.42]).

The contact deformations $d\mathbf{u}^{pq,\text{rigid}}$ and $\delta\theta^{pq,\,\text{rigid}}$ in equations [3.69]–[3.71] depend on the motions of the two particles p and q. These kinematic relationships are supplied by equations [2.2] and [2.18], which can be collected in matrix form as:

$$\begin{bmatrix} d\mathbf{u}^{pq,\text{rigid}} \\ d\theta^{pq,\,\text{rigid}} \end{bmatrix}_{6M\times 1} = [\,\mathbf{B}\,]_{6M\times 6N} \begin{bmatrix} d\mathbf{u} \\ d\theta \end{bmatrix}_{6N\times 1} \tag{3.72}$$

for all N particles and their M contacts. Matrix $[\mathbf{B}]$ is the kinematics matrix.

Equations [3.64], [3.65], [3.68], [3.71] and [3.72] are substituted into equation [3.60] to arrive at a matrix equation for all the particles in a granular assembly or cluster:

$$\left(\left[\,\mathbf{H}^{g-1}\,\right] + \left[\,\mathbf{H}^{g-2}\,\right] + \left[\,\mathbf{H}^{m}\,\right]\right) \left[\begin{array}{c}d\mathbf{u}\\ \hdotsfor{1} \\ d\theta\end{array}\right]_{6N\times1} = \left[\begin{array}{c}\delta\mathbf{b}\\ \hdotsfor{1} \\ \delta\mathbf{w}\end{array}\right]_{6N\times1} \qquad [3.73]$$

where the "mechanical" stiffness $[\mathbf{H}^{m}]$ is:

$$[\,\mathbf{H}^{m}\,]_{6N\times6N} = -\,[\,\mathbf{A}_1\,]_{6N\times2(6M)} \left[\begin{array}{c}\mathbf{F}\\ \hdotsfor{1} \\ \mathbf{M}\end{array}\right]_{2(6M)\times6M} [\,\mathbf{B}\,]_{6M\times6N} \qquad [3.74]$$

and the second geometric stiffness $[\mathbf{H}^{g-2}]$ is:

$$\left[\,\mathbf{H}^{g-2}\,\right]_{6N\times6N} = -\,[\,\mathbf{A}_1\,]_{6N\times2(6M)}\,[\,\mathbf{A}_2\,]_{2(6M)\times6N} \qquad [3.75]$$

with $[\mathbf{A}_2]$ given by equation [3.68]. This geometric stiffness accounts for the rotations of contact forces that accompany the rolling and twirling of particle pairs. The stiffness $[\mathbf{H}^{m}]$ in equation [3.74] is the conventional mechanical stiffness matrix for a system of N nodes that interact through M connections, but in a granular system, the connections are through contacts whose positions and orientations are altered by the particle movements – even by infinitesimal movements. The geometric alterations are captured, in part, with the matrices $[\mathbf{H}^{g-1}]$ and $[\mathbf{H}^{g-2}]$. A third alteration is also required.

To attain the desired form of equation [3.44], the co-rotating forces $\delta\mathbf{b}$ and $\delta\mathbf{w}$ on the right of equation [3.73] must be converted into the conventional increments $d\mathbf{b}$ and $d\mathbf{w}$, as would be seen by a global observer. In view of equations [3.55] and [3.56]:

$$\left[\begin{array}{c}d\mathbf{b}\\ \hdotsfor{1} \\ d\mathbf{w}\end{array}\right]_{6N\times1} = \left[\begin{array}{c}\delta\mathbf{b}\\ \hdotsfor{1} \\ \delta\mathbf{w}\end{array}\right]_{6N\times1} + \left[\,\mathbf{H}^{g-3}\,\right]_{6N\times6N} \left[\begin{array}{c}d\mathbf{u}\\ \hdotsfor{1} \\ d\theta\end{array}\right]_{6N\times1} \qquad [3.76]$$

where the third geometric stiffness $[\mathbf{H}^{g-3}]$ collects the relations in equations [3.48], [3.55] and [3.56] for all N particles:

$$\left.\begin{array}{l}d\theta^p \times \mathbf{b}^p = -d\theta^p \times \displaystyle\sum_{q\in Q^p} \mathbf{f}^{pq} \\[2ex] d\theta^p \times \\ \mathbf{w}^p = -d\theta^p \times \sum_{q\in Q^p}(\mathbf{r}^{pq} \times \mathbf{f}^{pq} + \mathbf{m}^{pq})\end{array}\right\} \rightsquigarrow \left[\,\mathbf{H}^{g-3}\,\right]_{6N\times6N}\left[\begin{array}{c}d\mathbf{u}\\ \hdotsfor{1} \\ d\theta\end{array}\right]_{6N\times1} \qquad [3.77]$$

The global stiffness matrix [H] in equation [3.50] embodies four stiffness components: two geometric components, $[\mathbf{H}^{g-1}]$ and $[\mathbf{H}^{g-2}]$, that depend on the particle shapes (surface curvatures) and on the current contact forces; a third geometric component $[\mathbf{H}^{g-3}]$ that depends on the particle size (the radial vectors $\mathbf{r}^{pq}$) as well as on the current contact forces; and a mechanical component $[\mathbf{H}^m]$ that depends on the contact stiffnesses. The combined geometric stiffness $[\mathbf{H}^g]$ is usually non-symmetric and its relative importance scales with the current contact forces, **f** and **m**. The mechanical stiffness can also be non-symmetric, for example as a result of contact slip, and its importance scales with the mechanical stiffnesses of the contacts.

Computation of the assembly stiffness [H] is greatly simplified when geometric effects are ignored, although certain internal instabilities may be nullified by this omission. The simplified problem leads to the following approximation of stiffness [H]:

$$[\,\mathbf{H}\,]_{6N \times 6M} \approx [\,\mathbf{H}^m\,]_{6N \times 6N} \approx [\,\mathbf{A}\,]_{6N \times 6N} \begin{bmatrix} \mathbf{F} \\ \cdots \\ \mathbf{M} \end{bmatrix}_{6M \times 6M} [\,\mathbf{B}\,]_{6M \times 6N} \qquad [3.78]$$

The contact stiffness matrix $[\mathbf{F/M}]_{6M \times 6M}$ consolidates the "pq" and "qp" parts of equation [3.71] by leaving only the "pq" part. The matrix [A] is the simplified statics matrix, which is simply the transpose of the kinematics matrix:

$$[\,\mathbf{A}\,] \approx [\,\mathbf{B}\,]^{\mathrm{T}} \qquad [3.79]$$

a relation that is based on the principle of virtual work and is widely encountered in matrix structural analysis.

A difficulty far greater than computing the geometric matrices is encountered, however, when multiple contacts are either sliding or in a state of incipient sliding. In this situation, we must determine the particular stiffness branch (elastic, micro-slip or slip) that applies to each of the potentially sliding contacts. Even when micro-slip is ignored, as in equation [2.42], each combination of the two branches (elastic and slip) among each of the M^{slip} potentially sliding contacts will result in a different global stiffness [H] and a different solution of equation [3.44]. That is, $2^{M^{\text{slip}}}$ different combinations, each with its own solution, are embodied in equation [3.44].

Most solutions will be invalid, as one or more of the assumed branches of the M^{slip} contacts will be inconsistent with the slip (or elastic) conditions that were assumed in arriving at the particular solution (e.g. the conditions in equation [2.42]). A similar situation is encountered with particles that are on the verge of touching and might come into firm contact with certain combinations of particle movements. This situation of incipient (grazing) contact results in additional potential stiffness matrices with their own solutions, further complicating the search for a compatible solution.

Rather than determining which contacts are slipping, Agnolin and Roux [AGN 07b] used a matrix approach to compute the bulk elastic stiffness of sphere assemblies, by assuming that all contacts were persistently elastic. For frictional systems in which particles can slip and separate, Shi [SHI 88] proposed an iterative process applied to a time-stepping DDA algorithm. In this approach, the condition of each contact at the start of a time step is initially assumed to be the same as that at the end of the previous step. With each iteration within a single time step, the displacements are solved with the corresponding stiffness matrix, and inconsistencies in the assumed conditions of the contacts are met with a change in the contact conditions and the stiffness matrix. Iteration proceeds until a consistent solution is found for the time step [KE 95, DOO 04]. This method can become computationally expensive, however, for a large granular system with a large number of sliding (or potentially sliding) contacts. Moreover, even though a consistent solution might be found, we must question the uniqueness of this solution and the possible bifurcation of solutions.

With matrix stiffness methods, various interpretations of failure – internal instability, loss of controllability, loss of uniqueness and negative second-order work – can be approached from the perspective of structural mechanics [BAŽ 91, KUH 06]. For example, a loss of uniqueness in the solution of equation [3.44] is the analog of bifurcation in continuum solutions associated with the constitutive stiffness tensor, which has been advanced as a possible explanation of shear bands [RIC 76]. The loss of internal stability in a continuum setting, which is signaled by a loss of ellipticity of the constitutive tensor, is akin to the emergence of negative eigenvalues of matrix [**K**]. Likewise, the loss of controllability applies to both continuum and discrete models, and results from non-singularity of the uncontrolled rows and columns of the constitutive tensor or of the matrix [**K**].

The method in this section is primarily intended for analyzing quasi-static systems, in which particles are nudged in increments $[d\mathbf{u}/d\theta]$ to new locations by changes in the external forces $[d\mathbf{b}/d\mathbf{w}]$ and by changes in the boundary conditions. The general DDA method is intended for dynamic systems and includes inertial and damping effects (matrices) in addition to the stiffness $[\mathbf{K}]$ described in this section (see [SHI 88, SHI 93, KE 95, DOO 04]).

3.4. Contact dynamics method

The contact dynamics (CD) method is a rigid-particle rigid-contact method for simulating the rearrangement of particles during loading of a granular assembly. The method, developed by Jean Jacques Moreau and Michel Jean in the 1980s and 1990s [MOR 94, JEA 95, JEA 99], embraces the non-smoothness of the unilateral constraint (zero penetration) between two rigid particles and the non-smooth constraint of the Coulomb limit of rigid-frictional contacts. Rather than using an explicit scheme that integrates particle accelerations and updates the particle velocities (as in the molecular dynamics, DEM, and matrix-based DDA approaches of sections 3.2 and 3.3), the CD method uses an implicit scheme to resolve the velocity jumps that occur during a time interval so that they are consistent with the unilateral and rigid-frictional constraints. Other non-smooth methods have been proposed, such as methods that use a nonlinear programming framework and treat the unilateral and rigid-frictional conditions as the inequality constraints of an objective functional [TZA 95]. The solution of such programming problems is quite difficult; however, due to the non-convex character of the constraints. The more pervasive CD method was originally developed for disk assemblies with inter-particle forces, but the method has also been applied to three-dimensional assemblies and to assemblies with inter-particle moments [EST 08, HUA 13].

Some of the difficulties that arise in DEM simulations are largely circumvented with the CD method: the large number of small time steps that must be used with the DEM for assemblies in which the bulk modulus is much greater than the bulk stress (see equations [3.28] and [3.31]) or the multiple stiffnesses that are presented by corner-edge and edge-face contacts between polyhedral particles. The CD methods also have limitations: the inability to account for the compliance of contacts, which is essential for simulating and measuring bulk stiffness, and difficulties that arise in simulating bonded particles, pore fluids and complex particle shapes, or in coupling a discrete model with a continuum FEM model. With the DEM method, the time step must be tuned to assure numeric stability; whereas, with CD models, the time step must be tuned to assure convergence before the next time increment is entered. Like the DEM method but unlike matrix methods, the CD approach can yield a solution to a problem but without the information necessary to diagnose the uniqueness or stability of the particular solution. Although these limitations make CD inappropriate for studying certain geomechanics problems (liquefaction, wave propagation, small-strain constitutive behavior, incremental stress probes, etc.), the method can be an efficient means for studying problems that are dominated by particle rearrangements rather than by the contact stiffnesses.

3.4.1. *Contact velocities*

The CD algorithm is based on the following problem: finding the differences (jumps) in the contact velocities within an assembly across a time interval dt. In the simplest problem of two-dimensional assemblies with normal and tangential contact forces but no contact moments, a *contact velocity* is the relative velocity of two particles p and q in the vicinity of their contact (or potential contact) c. This relative contact velocity is given as the rate $\mathbf{v}^c = d\mathbf{u}^{c,\text{rigid}}/dt$, with the increment $d\mathbf{u}^{c,\text{rigid}}$ defined in equation [2.2]. The velocity of contact c has the scalar normal and tangential components $v^{c,\text{n}}$ and $v^{c,\text{t}}$ (see equation [2.3]). Note that a positive normal velocity $v^{c,\text{n}}$ causes the particles to separate. The scalar components for all of the M contacts can be collected in a $2M \times 1$ stacked column vector $[\mathbf{v}^{\text{C,n}}/\mathbf{v}^{\text{C,t}}]$, where $\mathbf{v}^{\text{C,n}}$ is the list of all normal velocities and $\mathbf{v}^{\text{C,t}}$ is the corresponding list of tangential velocities. The Roman "C" denotes the full collection of "contact" quantities, whereas "c" denotes an individual contact. The rates of all M contacts within an assembly of N particles depend on the velocities (both translational and rotational) of the particles. The particle velocities and rotations are collected in a $3N \times 1$ column vector $[\mathbf{v}^{\text{P}}/\omega^{\text{P}}]$, where $\mathbf{v}^{\text{P}}$ is the $2N \times 1$ list of particle translational velocities, $\mathbf{v}^{\text{P}} = d\mathbf{u}^p/dt$; ω^{P} is the $N \times 1$ list of rotational velocities $\omega^p = d\theta^p/dt$; "P" denotes the full set of particles. The contact velocities and particle velocities of a two-dimensional system are kinematically related through the kinematics matrix $[\mathbf{B}]$:

$$\begin{bmatrix} \mathbf{v}^{C,n} \\ \hdashline \mathbf{v}^{C,t} \end{bmatrix}_{2M\times1} = [\,\mathbf{B}\,]_{2M\times3N} \begin{bmatrix} \mathbf{v}^{P} \\ \hdashline \omega^{P} \end{bmatrix}_{3N\times1} \qquad [3.80]$$

The $[\mathbf{B}]$ matrix for the two-dimensional problem is a simplification of the more general matrix in equation [3.72].

A net force and moment on a particle p results from any imbalance of the externally applied force $\mathbf{b}^p$, the externally applied moment $\mathbf{w}^p$ and the internal contact forces, as in equations [3.21] and [3.22] for the DEM method (see Figure 3.4). The force and moment imbalances on all N particles are arranged in matrix form as:

$$\begin{bmatrix} \overset{\star}{\mathbf{f}}{}^{P} \\ \hdashline \overset{\star}{\mathbf{w}}{}^{P} \end{bmatrix}_{3N\times1} = [\,\mathbf{A}\,]_{3N\times2M} \begin{bmatrix} \mathbf{f}^{C,n} \\ \hdashline \mathbf{f}^{C,t} \end{bmatrix}_{2M\times1} + \begin{bmatrix} \mathbf{b}^{P} \\ \hdashline \mathbf{w}^{P} \end{bmatrix}_{3N\times1} \qquad [3.81]$$

In this expression, the $2M$ normal and tangential contact forces are gathered into separate lists $\mathbf{f}^{C,n}$ and $\mathbf{f}^{C,t}$; the external forces and moments on the particles are gathered into the lists $\mathbf{b}^P$ and $\mathbf{w}^P$; the imbalances in the particle forces and moments are gathered into the lists $\overset{\star}{\mathbf{f}}{}^P$ and $\overset{\star}{\mathbf{w}}{}^P$. Note that contact moments (perhaps with notation $\mathbf{m}^C$) are excluded in this simplified presentation. To be consistent with the notations in section 2.2, the normal contact forces in the list $\mathbf{f}^{C,n}$ are the compressive normal forces $f^{c,n}$ at the individual contacts (see equation [2.22]). The matrix $[\mathbf{A}]$ is the statics matrix, similar to that in equation [3.61], and the statics matrix is the transpose of the kinematics matrix $[\mathbf{B}]$ (see equation [3.79]). For the simple two-dimensional problem, the moment lists, $\mathbf{w}^P$ and $\overset{\star}{\mathbf{w}}{}^P$, are lists of the counterclockwise scalar moments w_3 and $\overset{\star}{w}_3$ on the particles; the list of particle forces, $\mathbf{b}^P$ and $\overset{\star}{\mathbf{f}}{}^P$, can be separated into lists of their x_1 and x_2 components acting on the particles; the translations and rotations, the $\mathbf{v}^P$ and ω^P in equation [3.81], can be separated into lists of the v_1, v_2 and ω_3 components:

$$\begin{bmatrix} \overset{\star}{\mathbf{f}}{}^{P} \\ \hdashline \overset{\star}{\mathbf{w}}{}^{P} \end{bmatrix}_{3N\times1} = \begin{bmatrix} \overset{\star}{\mathbf{f}}{}^{P}_{1} \\ \hdashline \overset{\star}{\mathbf{f}}{}^{P}_{2} \\ \hdashline \overset{\star}{\mathbf{w}}{}^{P}_{3} \end{bmatrix} , \quad \begin{bmatrix} \mathbf{b}^{P} \\ \hdashline \mathbf{w}^{P} \end{bmatrix}_{3N\times1} = \begin{bmatrix} \mathbf{b}^{P}_{1} \\ \hdashline \mathbf{b}^{P}_{2} \\ \hdashline \mathbf{w}^{P}_{3} \end{bmatrix} , \quad \text{and} \quad \begin{bmatrix} \mathbf{v}^{P} \\ \hdashline \omega^{P} \end{bmatrix}_{3N\times1} = \begin{bmatrix} \mathbf{v}^{P}_{1} \\ \hdashline \mathbf{v}^{P}_{2} \\ \hdashline \omega^{P}_{3} \end{bmatrix} \qquad [3.82]$$

The differences (jumps) in particle velocities during the time step dt are given by a form of Newton's equation:

$$\left[\mathbf{M}^{\mathrm{P}}\right]_{3N\times3N}\left(\begin{bmatrix}\mathbf{v}^{\mathrm{P}}\\\hdashline\omega^{\mathrm{P}}\end{bmatrix}^{t+dt}-\begin{bmatrix}\mathbf{v}^{\mathrm{P}}\\\hdashline\omega^{\mathrm{P}}\end{bmatrix}^{t}\right)_{3N\times1}=dt\begin{bmatrix}\overset{\star}{\mathbf{f}}{}^{\mathrm{P}}\\\hdashline\overset{\star}{\mathbf{w}}{}^{\mathrm{P}}\end{bmatrix}_{3N\times1} \qquad [3.83]$$

The mass matrix $[\mathbf{M}^{\mathrm{P}}]$ is a diagonal matrix containing the masses and moments of inertia of the particles, and for the simple two-dimensional case, the mass matrix is a composite of three submatrices:

$$\left[\mathbf{M}^{\mathrm{P}}\right]_{3N\times3N}=\begin{bmatrix}\mathbf{M}^{\mathrm{P}}_{N\times N} & \mathbf{0} & \mathbf{0}\\\hdashline \mathbf{0} & \mathbf{M}^{\mathrm{P}}_{N\times N} & \mathbf{0}\\\hdashline \mathbf{0} & \mathbf{0} & \mathbf{I}^{\mathrm{P}}_{N\times N}\end{bmatrix} \qquad [3.84]$$

where elements of the $N \times N$ diagonal submatrices, mass $[\mathbf{M}^{\mathrm{P}}]$ and moment of inertia $[\mathbf{I}^{\mathrm{P}}]$, are arranged in the same order (of particles) as in the rows of equation [3.82]. Multiplying equation [3.83] by the inverse of $[\mathbf{M}^{\mathrm{P}}]$ gives:

$$\begin{bmatrix}\mathbf{v}^{\mathrm{P}}\\\hdashline\omega^{\mathrm{P}}\end{bmatrix}^{t+1}_{3N\times1}-\begin{bmatrix}\mathbf{v}^{\mathrm{P}}\\\hdashline\omega^{\mathrm{P}}\end{bmatrix}^{t}_{3N\times1}=dt\left[\mathbf{M}^{\mathrm{P}}\right]^{-1}_{3N\times3N}\begin{bmatrix}\overset{\star}{\mathbf{f}}{}^{\mathrm{P}}\\\hdashline\overset{\star}{\mathbf{w}}{}^{\mathrm{P}}\end{bmatrix}_{3N\times1} \qquad [3.85]$$

Equations [3.81] and [3.85] adjust the particle velocities at each time step, and the new velocities are then used to advance the particle positions. At this stage in its description, the CD approach appears similar to the DEM algorithm of section 3.2. Except for the damping terms in equations [3.23] and [3.26], the DEM equations are identical to equation [3.85]. It is at this point, however, that the CD method diverges from the DEM algorithm. In the discrete element method, the forces on particles are determined at time increment t, and an explicit integration approach is used to find the velocities at increment $t + 1/2$ and the positions at increment $t + 1$ (see equations [3.23]–[3.27]). With the contact dynamics method, equations [3.80], [3.81] and [3.85] are combined, placing focus on the contact velocities and contact forces [MOR 04, RAD 09,

UNG 03]:

$$\begin{bmatrix} \mathbf{v}^{C,n} \\ \cdots \\ \mathbf{v}^{C,t} \end{bmatrix}^{t+1}_{2M\times 1} - \begin{bmatrix} \mathbf{v}^{C,n} \\ \cdots \\ \mathbf{v}^{C,t} \end{bmatrix}^{t}_{2M\times 1}$$

$$= dt\, [\,\mathbf{B}\,]_{2M\times 3N}\, \big[\,\mathbf{M}^{P}\,\big]^{-1}_{3N\times 3N}\, [\,\mathbf{A}\,]_{3N\times 2M}\, \begin{bmatrix} \mathbf{f}^{C,n} \\ \cdots \\ \mathbf{f}^{C,t} \end{bmatrix}_{2M\times 1}$$

$$+ dt\, [\,\mathbf{B}\,]_{2M\times 3N}\, \big[\,\mathbf{M}^{P}\,\big]^{-1}_{3N\times 3N}\, \begin{bmatrix} \mathbf{b}^{P} \\ \cdots \\ \mathbf{w}^{P} \end{bmatrix}_{3N\times 1} \qquad [3.86]$$

With the DEM, contacts are compliant, and the forces at a contact, $f^{c,n}$ and $f^{c,t}$, depend on the contact displacement $d\mathbf{u}^{c,\mathrm{rigid}}$. This DEM approach is inconsistent with the hard-contact, unilateral (Signorini) condition of zero indentation, in which an infinitesimal contact indentation produces infinite force. The CD approach is instead based on the jumps in contact velocities that are consistent with hard-particle collisions, such that velocity jumps take the place of the smoother evolution of particle accelerations that are the basis of the DEM.

The jumps in contact velocities on the left of equation [3.86] depend on the contact forces on the right, so that the forces must be solved to find the velocities that satisfy the unilateral condition. For this reason, an implicit integration scheme is used to update the contact velocities, which are then used to resolve the particle velocities and advance the particle positions. The contact forces induce the velocity jumps that prevent inter-penetration of the particles and, thus, are necessary in satisfying the complementarity conditions at the contact, as described below.

One advantage of this approach applies to non-smooth particles, such as polygons or polyhedra, in which contact can occur in different combinations of corners, edges and faces: corner–corner contact, face–edge contact, corner–edge contact, etc. With the discrete element method, separate rules are required for the overlap–force relationship; whereas, the CD approach, with its zero-penetration basis, consistently enforces equation [3.86] for all categories of contact. As with the DEM, the CD approach can accommodate multiple contacts between particles, as can occur with non-convex particles.

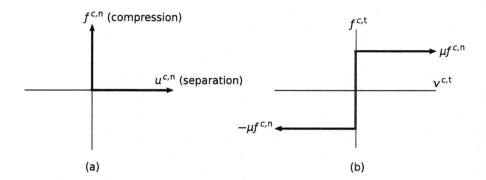

Figure 3.5. *Rigid particle contact conditions: (a) Signorini set for normal force and displacement; (b) Coulomb friction set for tangential force and displacement*

3.4.2. *Complementarity relations*

The assumption of rigid, impenetrable particles imposes constraints on the normal and tangential forces and velocities. The contact indentation ζ of equation [2.24] must be non-positive, and the normal negative-overlap $u^{c,n} = -2\zeta$ of equation [2.3$_1$] must be non-negative. When two rigid particles are touching at contact c, with zero overlap, the normal compressive force $f^{c,n}$ is non-negative but indeterminate. This Signorini condition is expressed as [MOR 94]:

$$u^{c,n} > 0 \Rightarrow f^{c,n} = 0 \tag{3.87}$$

$$u^{c,n} = 0 \Rightarrow f^{c,n} \geq 0$$

as shown in Figure 3.5(a). Note again that a positive normal force $f^{c,n}$ is compressive. The set of admissible displacements and forces in equation [3.5] can also be specified with the combination of a linear and a complementarity constraint: $u^{c,n} + f^{c,n} > 0$ and $u^{c,n} f^{c,n} = 0$. In either formulation, the set of admissible displacements and forces is non-convex, which presents an obstacle to solving the movements with standard linear or quadratic programming methods. Moreover, when two particles are touching, the normal contact velocity $v^{c,n}$ must be either zero (for a persistent contact) or greater than zero (for a separating contact), and a non-zero normal force is only consistent with persistent contact, leading to the following three admissible conditions:

$$u^{c,n} > 0 \qquad\qquad \Rightarrow f^{c,n} = 0$$
$$u^{c,n} = 0 \text{ and } v^{c,n} > 0 \quad \Rightarrow f^{c,n} = 0 \qquad\qquad [3.88]$$
$$u^{c,n} = 0 \text{ and } v^{c,n} = 0 \quad \Rightarrow f^{c,n} \geq 0$$

The tangential force $f^{c,t}$ and velocity $v^{c,t}$ of two touching rigid particles must comply with the rigid-frictional Coulomb constraint [TZA 95]:

$$v^{c,t} < 0 \Rightarrow f^{c,t} = -\mu f^{c,n}$$
$$v^{c,t} = 0 \Rightarrow f^{c,t} \in [-\mu f^{c,n}, \mu f^{c,n}] \qquad\qquad [3.89]$$
$$v^{c,t} > 0 \Rightarrow f^{c,t} = \mu f^{c,n}$$

as shown in Figure 3.5(b). Unlike the DEM, the tangential force depends only on the current contact velocity and does not involve the history of contact interactions.

3.4.3. Contact forces

In section 3.3, the stiffness matrix was derived for an assembly of particles with compliant, non-rigid contacts. The movement of one particle affects the equilibrium of all other particles, so that a large set of simultaneous equations must be solved to find the compatible movements of all particles. Rather than directly confronting the difficulty of assuring the equilibrium of all particles, the discrete element method (DEM) considers the local equilibrium of single particles and the imbalance of forces produced by each particle's immediate neighbors. The DEM algorithm uses explicit dynamic relaxation to bring all particles into near-equilibrium (equations [3.21]–[3.26]). The contact dynamics (CD) approach faces a similar difficulty, by requiring that the entire set of $2M$ simultaneous equations [3.86] is satisfied while also meeting the Signorini and Coulomb constraints at every contact. Rather than directly confronting this difficult computational problem, the CD algorithm considers the velocity–force relation at individual contacts and uses an iteration scheme to assure that the particles' velocities are compatible with nearly all contact constraints.

Equation [3.86] can be reduced to the following form, which gives the contact velocities at $t + 1$ in terms of the velocities at t and the contact and external forces:

$$
\begin{bmatrix} \mathbf{v}^{C,n} \\ \hline \mathbf{v}^{C,t} \end{bmatrix}_{2M\times1}^{t+1} = \begin{bmatrix} \mathbf{v}^{C,n} \\ \hline \mathbf{v}^{C,t} \end{bmatrix}_{2M\times1}^{t}
\tag{3.90}
$$

$$
+ \; [\, \mathbf{Q}_1 \,]_{2M\times2M} \begin{bmatrix} \mathbf{f}^{C,n} \\ \hline \mathbf{f}^{C,t} \end{bmatrix}_{2M\times1} + \; [\, \mathbf{Q}_2 \,]_{2M\times3N} \begin{bmatrix} \mathbf{b}^{P} \\ \hline \mathbf{w}^{P} \end{bmatrix}_{3N\times1}
$$

where matrices $[\mathbf{Q}_1]$ and $[\mathbf{Q}_2]$ represent the corresponding products in equation [3.86]. Taking matters a step further, the normal and tangential velocities of a *single* contact c at increment $t + 1$ can be expressed in terms of the unknown forces at the same contact [RAD 09]:

$$
\begin{bmatrix} v^{c,n} \\ v^{c,t} \end{bmatrix}_{2\times1}^{t+1} = \begin{bmatrix} \mathbf{C}^c \end{bmatrix}_{2\times2} \begin{bmatrix} f^{c,n} \\ f^{c,t} \end{bmatrix}_{2\times1}^{t+1} + \begin{bmatrix} a^{c,n} \\ a^{c,t} \end{bmatrix}_{2\times1}^{t}
\tag{3.91}
$$

The four elements in matrix $[\mathbf{C}^c]$ are simply the coefficients of $f^{c,n}$ and $f^{c,t}$ in $[\mathbf{Q}_1]$, but the vector $[\mathbf{a}]^t$ draws on the following information:

1) the known values of $v^{c,n}$ and $v^{c,t}$ at increment t;

2) the external forces $\mathbf{b}$ and $\mathbf{w}$ that act on the two particles that meet at the contact c;

3) the contact forces $\mathbf{f}^n$ and $\mathbf{f}^t$ at contacts other than c that also act on the same two particles. In using equation [3.91] to calculate the new forces for contact c, the other contact forces are assumed to retain their most recent values (at increment t or the updated values found during subsequent iterations). The products of these other contact forces with elements of $[\mathbf{Q}_1]$ contribute to the vector $[\mathbf{a}]^t$.

After computing $a^{c,n}$, $a^{c,t}$ and the four values in matrix $[\mathbf{C}^c]$, equation [3.91] becomes a system of two linear equations with four unknowns. These unknown values are the new, $t + 1$, velocities and forces $v^{c,n}$, $v^{c,t}$, $f^{c,n}$ and $f^{c,t}$, for the contact c. The additional information necessary to solve these unknowns is given in the complementarity conditions of equations [3.88] and [3.89].

The contact velocities and forces are found in a similar manner for all M contacts within the assembly, recognizing that when solving for a particular contact c, equation [3.91] treats the forces at other contacts as if they are known quantities. The velocities and forces at all M contacts are not solved simultaneously, but are solved one-by-one, as in a Gauss–Seidel approach. The CD algorithm requires a repeated application of equation [3.91] to the M contacts until the contact forces and velocities converge to stable values. With each iteration, the values of the contact forces from the previous iteration are used to update the forces. As with the DEM approach, there is no assurance that the values at $t + 1$ are unique solutions of the global problem (i.e. equations [3.86], [3.88] and [3.89] in the case of CD).

After finding a stable solution at $t + 1$, the particle positions are advanced by applying equations [3.81] and [3.85] with the newly determined contact forces. A search for new contacts, potential contacts and separated contacts is then undertaken before undertaking the next time step.

Loading, Movement, and Strength

4.1. Deformation and strength

Laboratory tests of granular soil specimens are routinely conducted to determine the stiffness, deformation and strength characteristics of these materials, with the intent of applying the results to larger field situations. The most useful tests are those that maintain nearly uniform deformation and stress within the specimen, so that the measured relationship of stress and strain represents the bulk material behavior rather than the behavior of a local region within the test specimen. Laboratory tests should also measure or control the pore fluid pressure, so that the effective stress can be determined. Testing methods that satisfy these conditions include uniaxial compression (consolidation or oedometer) tests, triaxial compression and extension tests, and plane-strain biaxial compression tests. Uniaxial compression tests allow the control of a single quantity: either stress or strain in the axial direction. Traditional triaxial and plane-strain biaxial conditions allow the control of two components chosen from the axial strain ε_{11}, axial stress σ_{11}, lateral strain ε_{33} ($= \varepsilon_{22}$ for triaxial tests) and lateral stress σ_{33} ($= \sigma_{22}$ for triaxial tests that employ an outer rubber membrane or $\varepsilon_{22} = 0$ for plane-strain biaxial tests). The remaining, uncontrolled quantities are measured, representing the material's response to the loading, and this response is analyzed to gain insight into the constitutive behavior. Shear stresses are ideally minimized along the boundaries of a triaxial or plane-strain biaxial specimen, so the controlled and responding stresses and strains are the major and minor principal components. Traditional triaxial testing permits either of two conditions for the intermediate principal stress: during axial compression, the specimen is shortened and the intermediate stress matches the smaller (minor) compressive stress; but when the specimen is axially extended, the intermediate stress equals the larger (major) compressive stress. More advance testing methods, such as with true-triaxial, hollow cylinder and cylindrical torsion apparatus, enable control of additional stress and strain components and are used for measuring the effects of the intermediate principal stress and strain. Most testing apparatuses allow either measurement or control of the internal pore fluid pressure, as well as the externally applied strain and stress (i.e. the total stress). In the following, stress is measured relative to the pore water pressure, which becomes, in essence, a datum isotropic stress, so that stress σ is the effective, inter-granular stress (see equation [1.48] and its discussion). Laboratory testing is typically conducted under either drained or undrained conditions, because these two conditions are the easiest to enforce, even though they comprise only the extremities of a range of volumetric conditions. During drained loading, the pore pressure datum is maintained constant, whereas the volume is maintained constant under undrained conditions. Undrained conditions are typically achieved with saturated specimens, by preventing the flow of water into or out of the specimen.

Typical trends are shown in Figure 4.1 for triaxial compression tests on uncemented sands. In the figure, p is the mean compressive effective stress (negative mean stress); p_0 and e_0 are the initial negative mean stress and void ratio; the compressive deviator stress q is the difference between the major and minor principal stresses, $q = -(\sigma_1 - \sigma_3)$; ε is a measure of compressive deviatoric strain; v is the volumetric strain with expansion being positive. Although behavior is nearly elastic at the start of loading, plastic deformation, although slight, begins to occur at strains ε of as small as 1×10^{-5}, and the elasto-plastic stiffness is steadily degraded during subsequent strain-hardening until the material eventually fails. The response to loading is seen to depend on the initial void ratio and the initial mean stress. Dense materials or materials with low initial mean stress tend to behave in a more dilative manner, in which the volume tends to increase under drained conditions or the mean effective stress tends to increase with undrained conditions. The opposite tendency – contractive behavior – applies to loose (low density) materials or during tests with a high initial mean stress. The figure indicates that the ratio q/p, a dimensionless measure of strength, eventually attains a steady, constant residual value M at large strains, a condition that is characteristic of the *critical state*. The eventual ratio is insensitive to the mean stress and to the initial void ratio. Under drained conditions, a higher *peak* ratio can be reached at medium strains, but the peak ratio depends on the initial density and mean stress and is larger for dense materials or for materials that are tested with a small initial mean stress. For dense soils, softening (i.e. reduction in the deviatoric stress) occurs at strains beyond the peak state and until the critical state is reached. With very loose sands in drained loading, the peak condition is only reached at large strains, so that the peak and critical state ratios q/p are equal. Strength is often expressed as an angle of internal friction ϕ, which can be applied to the peak strength or to the eventual residual condition. The angle is related to the ratio of principal effective stresses $m = \sigma_1/\sigma_3$ or to the ratio $M = q/p$ that is attained at large strains. For triaxial loading, angle ϕ is given by:

$$\phi = \sin^{-1}\left(\frac{m-1}{m+1}\right) = \begin{cases} \sin^{-1}\left(3M/(6+M)\right) & \text{compression, } \sigma_2 = \sigma_3 \\ \sin^{-1}\left(3M/(6-M)\right) & \text{extension, } \sigma_1 = \sigma_2 \end{cases} \quad [4.1]$$

noting that $m \geq 1$ and $M \geq 0$. With two-dimensional assemblies, the deviator and mean stresses are $q = -(\sigma_1 - \sigma_2)$ and $p = -\frac{1}{2}(\sigma_1 + \sigma_2)$, and equation [4.1] is replaced with:

$$\phi_{2D} = \sin^{-1}\left(\frac{m-1}{m+1}\right) = \sin^{-1}(M/2) \quad [4.2]$$

Subscripts can be used to denote the angles corresponding to the peak or critical states, ϕ_{peak} or ϕ_{CS}, where the peak ratio $(q/p)_{\text{peak}}$ is used to compute the former and M is used for the latter. The relative rate of volume change is often expressed as the dilatancy angle ψ, computed from the rates of the principal strains:

$$\sin\psi = -\frac{\varepsilon_1 + \varepsilon_3}{\varepsilon_1 - \varepsilon_3} \quad [4.3]$$

where ε_1 and ε_3 are the major and minor extensional strains (the principal strain ε_3 is replaced with ε_2 for two-dimensional assemblies). As seen in Figure 4.1(a), the dilation rate is usually negative at the start of drained loading, but the dilatancy angle ψ transitions to a positive value for all but very loose specimens. Strength and the rate of volume change depend on the loading conditions: in particular, on the value of the intermediate principal stress σ_2 in relation to the major and minor principal stresses, σ_1 and σ_3. Behavior also depends on whether the directions of the principal strains, ε_1, ε_2 and ε_3, are unchanged or rotate during loading.

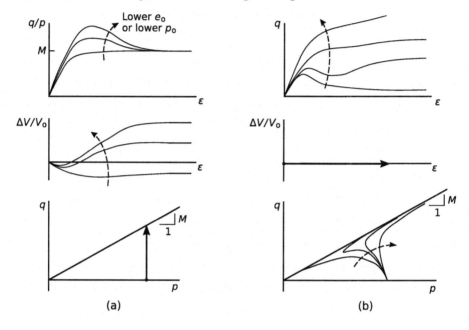

Figure 4.1. *Behavior of granular specimens in traditional triaxial compression tests: (a) drained loading; (b) undrained compression loading*

Both dense and loose specimens eventually reach a condition of stationary density, at which $\psi = 0$, but dense specimens will dilate more and loose specimens will dilate less before attaining this final condition. This observation is part of the critical state concept, which holds that specimens of a particular soil, whether loose or dense, eventually reach a common density at large deviatoric strain [SCH 68]. Moreover, after sustained shearing from an initial packing, a granular material reaches a steady condition (the critical state) in which the ratio of deviatoric stress to mean stress remains nearly stationary during further shearing. The critical state is distinguished, therefore, by its stationary and convergent qualities. Not only does the stress converge to a steady condition, but also other bulk measures of internal fabric and structure – void ratio, coordination number, fabric anisotropy, contact force distributions, etc. – converge and remain stationary as well

[THO 00, PEÑ 09]. Although the values of these characteristics will depend on the mean stress, the material's composition (particle shape, size distribution, texture of particle surfaces, etc.) and the form of loading (mean stress, intermediate principal stress and rate of principal stress rotation), for a given material, mean stress, and type of loading, a specific steady condition is eventually attained. The author proposed a third characteristic of the critical state: at the critical state, the micro-scale landscape of particle arrangements and particle interactions exhibits the greatest diversity and disorder that is consistent with information known *a priori* – information, such as local equilibrium and kinematic compatibility, which constrains the available landscape of particle configurations and interactions [KUH 16b].

For triaxial compression loading, the ratio $M = q/p$ at the critical state is nearly independent of p, but as was noted, the bulk internal fabric and structure, including void ratio, that are reached at the critical state will depend on the mean stress. Lower void ratios and higher coordination numbers are attained with increasing mean stress. The unique relationship between the critical state void ratio and the mean effective stress is expressed as a *critical state line* in the space of density and mean stress. A material that is initially more dense than the critical state density that corresponds to its initial mean stress is said to be in an initial *dilative state*; materials that are less dense than the critical state density for the initial stress are said to be in an initial *contractive state*. The difference between the initial and critical state densities is a state parameter can aid in predicting the subsequent loading behavior [BEE 85].

The loading of a granular material produces profound changes in its internal fabric and structure, as is seen in the remainder of this chapter. These changes are overtly expressed through fabric alterations, such as the number of contacts and the orientations and arrangements of particles; but changes also occur as subtle redistributions of the contact forces and adjustments in the normal and sliding components of these forces. Changes in the fabric and load-bearing structure usually produce a progressive degradation of bulk stiffness that accompanies an increase of the loading strain, and the bulk stiffness moduli also become anisotropic during loading [KUH 15]. The initial elastic moduli are reduced with even the smallest loading strain (strains ε as small as 1×10^{-5}), and further loading is attended by both reversible and irreversible (elastic and plastic) deformations, with the latter growing in prominence until the bulk deformation is entirely irreversible at the critical state. A degradation of the bulk stiffness is usually accompanied by a growing anisotropy of the bulk elastic moduli [KUH 15].

Prior to the critical state, irreversible deformation can be much larger than the reversible strains, which can lead to alternative understandings of the very meaning of failure in granular materials: loss of control, loss of uniqueness (bifurcation), negative second-order work, etc. From a practical point of view, certain components of stress and strain (or their linear combinations of these quantities) are controlled during a laboratory test or in the field, and the loss of this control (i.e. loss of controllability) usually constitutes a failure condition: when the inner product of the controlled quantities and their complementary uncontrolled quantities vanish [NOV 94]. Failure can be either localized or diffuse: with the former, failure is accompanied by the emergence of shear bands, at or near the peak state, whereas diffuse failure occurs more uniformly throughout the material.

4.1.1. *Two phases of deformation*

DEM simulations of quasi-static monotonic loading usually display erratic fluctuations of the bulk deviatoric stress. This behavior is shown in Figure 4.2(a) as the irregular and raspy changes in the normalized deviator stress q/p for a simulation of constant-p triaxial compression of an assembly of sphere-clusters. Without these fluctuations, the behavior in Figure 4.2(a) is similar to the smooth middle line in the stress–strain plot at the top of Figure 4.1(a). The fluctuations, which are absent at the smallest strains, become most notable after the peak stress is attained, although they also occur at smaller, pre-peak strains during the strain hardening phase of loading. The magnitude and frequency of the stress fluctuations depend on the assembly size: their magnitudes decrease and the frequency increases as the size of the assembly (its number of particles) is made larger (see Figure 4.2(b)) [KUH 09]. Stress fluctuations are observed in laboratory tests of sands and in laboratory models of rods and disks, although the fluctuation magnitudes in laboratory tests on sands are typically small due to the large number of particles that are involved. The fluctuations are usually larger for specimens of spherical particles (for example, glass ballotini) than for sands (see [ALS 06]). Roux and Combe [ROU 02, ROU 03] have suggested that these stress fluctuations result from sudden rearrangements of the particles, and they have divided granular behavior into two regimes of deformation: a *strictly quasi-static regime* at very small strain and a *rearrangement regime* at larger strains, where the abrupt stress fluctuations are exhibited. The strain separating the two regimes depends on the relative stiffness of the contacts: the threshold strain being smaller for relatively stiff contacts. In the author's DEM simulations of non-convex sphere-clusters in which contact stiffness is similar to that of sand particles, the threshold strain is in the range of 0.1–0.2%.

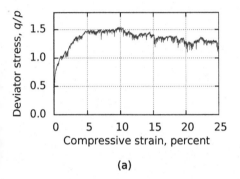
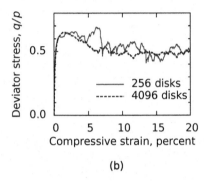

(a)   (b)

Figure 4.2. *DEM simulation results of the normalized deviator stress q/p versus compressive strain: (a) triaxial compression of an assembly of 10,000 sphere-clusters; (b) biaxial compression of disk assemblies*

In the strictly quasi-static regime of smooth stress–strain behavior, the bulk material stiffness is primarily influenced by deformations at the contacts, even though some contact slipping can occur at these small strains. The gradual reduction of stiffness that occurs with increasing strain is mainly due to a diminishing number of contacts, a tendency discussed in section 4.3.2. Because of the dominant influence of contact deformation and stiffness at small strains, the strictly quasi-static regime is best simulated with numerical methods that employ non-rigid contact models (the DEM and stiffness matrix methods in sections 3.2 and 3.3), rather than methods that enforce strictly hard contacts, such as the contact dynamics (CD) method (section 3.4). As behavior transitions into the rearrangement regime, a greater fraction of contacts reach the friction limit and are sliding, and the internal structure of the material becomes more fragile: the added slip or disengagement of individual contacts can produce local instabilities within the assembly, a reduction of statical redundancy, a rearrangement of the local contact network and a reduction of stress within the assembly.

Deformation within the rearrangement regime proceeds with frequent and seemingly abrupt alterations of stress – usually reductions in stress – caused by particle rearrangements. The sudden rearrangements of particles are evidenced by volatile changes in a specimen's volume during loading. These rearrangements are precipitated by the local loss of stability within groups of particles, producing negative second-order work at individual contacts [DAO 13]. The nature of instability within a granular material and its relation to the material's discrete stiffness are discussed near the end of section 3.3, but the fact that the magnitudes of stress fluctuations decrease with increasing assembly size indicates that the fluctuations result from localized particle rearrangements. By measuring the velocities of particles in DEM simulations, Peters and Walizer [PET 13] found that stress drops were accompanied by bursts in the kinetic energy of the particles, fed by reductions in the elastic energy at the contacts. In addition to bursts of kinetic energy, Thornton [THO 00] noted wide fluctuations in the number of contacts, indicating substantial rearrangement of the contact network. The kinetic nature of these events suggests that the manner in which particles are rearranged depends on the particles' inertias during the brief transfer of elastic energy to kinetic energy among the particles. These kinetics, which operate at the scale of individual particles, render the results of DEM simulations sensitive to the loading rate (strain increment) and to the amount of viscous damping that is ascribed to the particles.

Figure 4.3 shows details of stress and fabric during the rearrangement regime in simulations of non-convex sphere-cluster particles with Hertz-type contacts. The simulations are of triaxial compression at constant mean stress, also shown in Figure 4.2(a). Figure 4.3 details the behavior during compressive strains in the range of 0.61–0.66% for very slow monotonic loading. The figure encompasses two hundred thousand strain increments. The strains in Figure 4.3 occur well before the peak stress and correspond to the strain-hardening at the far left side of Figure 4.2. The fluctuation events typically occur within an average strain span of 0.0002% from peak to trough (for the small strain increments used in these simulations, the average stress drop spans 1000 strain increments). The finite durations of these intervals indicate that these stress drops are not entirely abrupt but have a duration that depends on the strain required for particles to find new and more stable arrangements. The stress drops are typically followed by a gradual increase in stress during which the stiffness (i.e. slope) progressively decreases until the next stress drop occurs. Figures 4.3(b) and (c) show the average coordination number and the fraction of sliding contacts. The stress drops are accompanied by a coincident reduction in the number of contacting particles and in the number of sliding contacts. The relatively large changes in these two measures of fabric indicate that significant adjustments in the particles' arrangement are associated with even the modest stress drops shown in the figure, which occurred during the strain-hardening phase of loading. Such is the complexity of deformation and failure in granular materials.

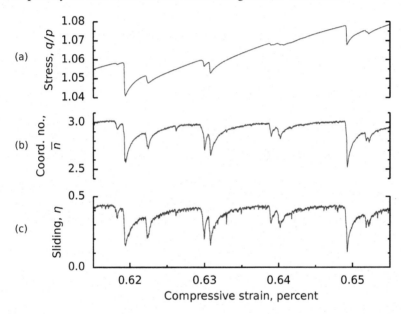

Figure 4.3. *Details of a DEM simulation, showing stress and fabric during a small interval of strain within the rearrangement regime of behavior. Multiple stress drops are shown in the following respects: (a) normalized deviator stress q; (b) average coordination number n̄; (c) fraction of sliding contacts relative to the number of particles.*

4.2. Trends in movements and rotations

The deformation of a bulk granular material is produced by the movements and rotations of its particles, by the relative movements among the particles at their contacts and by meso-scale deformation of particle clusters. This section concerns general trends in these local movements and rotations of particles and of particle groups during deformation. The spatial organization of these movements is the subject of section 4.7.

4.2.1. *Affine motion*

Simulations and model experiments have greatly contributed to our understanding of the micro- and meso-scale movements that occur during bulk deformation. Although movement data can, in many cases, be diverse, ambiguous and contradictory, they are best understood when compared with movements that would conform to a uniform deformation, *affine* field. An assumption of such conformity is also termed the *kinematic assumption* [LIA 97]. If the current configuration at time t is adopted as the reference configuration, the "ideal", affine translation $d\mathbf{u}^{p,\text{ideal}}$ of particle p and its ideal rotation $d\theta^{p,\text{ideal}}$, both during the increment dt, are:

$$d\mathbf{u}^{p,\text{ideal}} = \overline{\mathbf{L}} \cdot \mathbf{x}^p \, dt \quad \text{and} \quad d\theta^{p,\text{ideal}} = \tfrac{1}{2}(\overline{\mathbf{L}} - \overline{\mathbf{L}}^{\mathrm{T}}) \, dt = \overline{\mathbf{W}} \, dt \qquad [4.4]$$

where $\mathbf{x}^p$ is the position of the reference point attached to the particle (perhaps at its center), $\overline{\mathbf{L}}$ is the uniform mean velocity gradient of a granular region that includes p and $\overline{\mathbf{W}}$ is the uniform mean spin tensor (equations [1.93]–[1.94]). The ideal movement of the reference point $\mathbf{x}^q$ of a neighboring particle q relative to that of p is:

$$d\mathbf{u}^{pq,\text{ideal}} = \overline{\mathbf{L}} \cdot (\mathbf{x}^q - \mathbf{x}^p) \, dt = \overline{\mathbf{L}} \cdot \mathbf{l}^{pq} \, dt \qquad [4.5]$$

and the ideal relative movement $d\mathbf{u}^{c,\text{ideal}}$ of material points in p and q at a contact c is:

$$d\mathbf{u}^{c,\text{ideal}} = \tfrac{1}{2}(\overline{\mathbf{L}} + \overline{\mathbf{L}}^{\mathrm{T}}) = \overline{\mathbf{D}} \cdot \mathbf{l}^{pq} \, dt \qquad [4.6]$$

where $\overline{\mathbf{D}}$ is the symmetric rate of deformation tensor (equation [1.94$_1$]) and $\mathbf{l}^{pq}$ is the branch vector joining the reference points attached to particles p and q (Figure 1.4). Equation [4.5] is the counterpart of equation [2.1] when the movements $d\mathbf{u}^p$ and $d\mathbf{u}^q$ conform to the deformation $\overline{\mathbf{L}}$. The relative contact movement $d\mathbf{u}^{c,\text{ideal}}$ in equation [4.6] applies when the two particles' rotations in equation [2.2], $d\theta^p$ and $d\theta^q$, are also in accord with the underlying uniform rotation field – a rotation at the rate $\overline{\mathbf{W}}$. Under these conditions, the ideal relative movement at the contact is unaffected by the underlying uniform rotation, a result that is derived from equations [4.4] and [2.2]. Movements $d\mathbf{u}^{pq,\text{ideal}}$ and $d\mathbf{u}^{c,\text{ideal}}$ can be separated into components that are normal to and parallel with the tangent plane at c, in the manner of equation [2.3]:

$$d\mathbf{u}^{pq,\text{ideal,n}} = d\mathbf{u}^{pq,\text{ideal}} \cdot \mathbf{n}^c \quad \text{and} \quad d\mathbf{u}^{pq,\text{ideal,t}} = d\mathbf{u}^{pq,\text{ideal}} - d\mathbf{u}^{pq,\text{ideal,n}}\mathbf{n}^c \quad [4.7]$$

$$d\mathbf{u}^{c,\text{ideal,n}} = d\mathbf{u}^{c,\text{ideal}} \cdot \mathbf{n}^c \quad \text{and} \quad d\mathbf{u}^{c,\text{ideal,t}} = d\mathbf{u}^{c,\text{ideal}} - d\mathbf{u}^{c,\text{ideal,n}}\mathbf{n}^c \quad [4.8]$$

In the alternative setting of a micro-polar continuum, the particle rotations are assumed to be influenced by a separate kinematic field, a micro-spin field $\overline{\mathbf{W}}^{\text{micro}}$ having a uniform spatial gradient $\nabla\overline{\mathbf{W}}^{\text{micro}}$. That is, the uniform kinematic field $\overline{\mathbf{L}}$ is augmented with an independent field of micro-rotation. If the uniform spin $\overline{\mathbf{W}}$ and micro-spin $\overline{\mathbf{W}}^{\text{micro}}$ are expressed as vorticity vectors $\overline{\omega}$ and $\overline{\omega}^{\text{micro}}$ (see equation [1.95]), and the uniform micro-spin gradient is expressed as the rank-two vorticity gradient $\overline{\mathbf{\Omega}}^{\text{micro}}$, such that $\overline{\Omega}_{ij}^{\text{micro}} = \partial\overline{\omega}_i^{\text{micro}}/\partial x_j$, then the ideal relative contact movement $d\mathbf{u}^{c,\text{ideal}}$ is:

$$d\mathbf{u}^{c,\text{micro-polar}} = \overline{\mathbf{D}} \cdot \mathbf{l}^{pq}\, dt + \left(\overline{\omega} - \overline{\omega}^{\text{micro}} - \overline{\mathbf{\Omega}}^{\text{micro}} \cdot \mathbf{r}^{c,p}\right) \times \mathbf{l}^{pq}\, dt \quad [4.9]$$

$$- (\overline{\mathbf{\Omega}}^{\text{micro}} \cdot \mathbf{r}^{c,q}) \times \mathbf{r}^{c,q}\, dt$$

where $\mathbf{r}^{c,p}$ and $\mathbf{r}^{c,q}$ are the contact vectors (Figure 1.4). Note that when the particles rotate in accord with the underlying uniform displacement field, such that $\overline{\omega} = \overline{\omega}^{\text{micro}}$ and $\overline{\mathbf{\Omega}}^{\text{micro}} = \mathbf{0}$, the ideal relative movement $d\mathbf{u}^{c,\text{micro-polar}}$ of the two particles at c is the same as that in equation [4.6].

4.2.2. Deviations from affine motion

A signature characteristic of granular materials is the large deviation (fluctuation) of particle movements from their ideal, affine movements – a characteristic that has confounded the development of accurate constitutive formulations based on simple kinematic assumptions. These non-affine displacements are exposed in the trajectory of the single particle shown in Figure 4.4. The trajectory is from a three-dimensional simulation of 4096 densely packed poly-disperse frictional spheres. The assembly was quasi-statically deformed in horizontal simple shear, and the trajectory is of the cumulative movement between the start of loading and a shear strain of

160%, which takes the material into the critical state. The uniform, ideal background movement $d\mathbf{u}^{p,\text{ideal}}$ of equation [4.4] was subtracted from the actual movement $d\mathbf{u}^p$ of the single particle. The particle size is about 1.0 mm and the assembly size is about 13 mm, so that its fluctuations are quite large when compared with either the particle size or the assembly size, with the accumulated drift being much larger in the direction of shearing, x_1 than in the transverse directions, x_2 and x_3. The erratic nature of the movement resembles that of a random walk (Wiener process). Although an ideal random walk, in which future steps are unaffected by the past, leads to a cumulative drift distance that increases with the square root of time (or of shearing strain, in the case of a uniform rate of deformation), the results in Figure 4.4 and those analyzed by Radjai *et al.* [RAD 02] indicate that the drift is superdiffusive, with a drift distance proportional to (strain)$^{0.9}$. In their simple-shear experiments on wood disk assemblies, Richefeu *et al.* [RIC 12] also found that the average drift distance of a particle from its ideal, affine position was superdiffusive and increased almost linearly with strain.

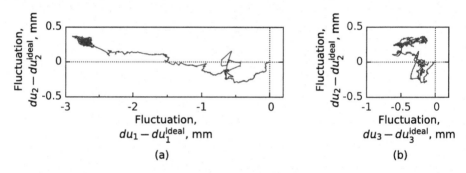

Figure 4.4. *Displacement fluctuations relative to the ideal affine displacements (i.e. $d\mathbf{u}^p - d\mathbf{u}^{p,\text{ideal}}$) of a single sphere during horizontal simple shearing deformation in the x_1 direction across the x_2 height: (a) fluctuations in the shearing plane; (b) fluctuations in the transverse plane*

Displacement fluctuations, $d\mathbf{u}^p - d\mathbf{u}^{p,\text{ideal}}$, are typically smaller at the start of loading than they are at the peak or critical states [ROU 03]. The probability densities in Figure 4.5 are of displacement fluctuations determined in simulations of drained triaxial compression of two assemblies: one of spheres and the other of sphere clusters (also see [GUO 14]). The figure shows the non-affine displacements at the start of loading and at the critical state. The incremental movements in this figure are normalized by dividing by the small increment of compressive strain and by the median particle size. At the critical state (Figure 4.5(b)), the non-affine movements are substantial and occur at rates several times greater than the affine part of movement, $du^{p,\text{ideal}}$. Although at the start of loading (Figure 4.5(a)), the non-affine movements are considerably smaller than those at the critical state, even at 0% strain, a number of particles had fluctuations with magnitudes greater than the mean affine movement.

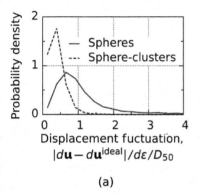 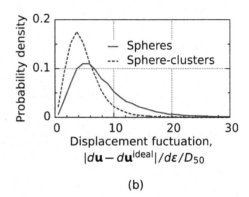

(a)   (b)

Figure 4.5. *Probability density of displacement fluctuations of drained triaxial compression. (a) at the start of loading; (b) at the critical state. The displacements were measured across a compressive strain interval of 2.5 × 10^{-5}*

The relative movements of the centers of contacting particles are denoted as du^{pq} in equation [2.1]. During slow, quasi-static deformation of dense granular assemblies, the actual relative movements between particles differ greatly from the ideal relative movements $du^{pq,\text{ideal}}$, as each particle must adjust its position to maintain equilibrium within its neighborhood of particles. An appreciation of the disparity between the actual and ideal, affine movements can be gained from Figure 4.6 (also see [KUH 16a]). The figure gives the results of a DEM simulation of triaxial compression of a sphere assembly. The results are at the compressive strain of 0.01%, near the start of loading. Each dot in Figure 4.6(a) represents the *tangential* component of the relative movement, $du^{pq,t}$, of a single pair of contacting particles (as the tangential component of equation [2.1]). The tangent planes have a different orientation for each pair, but all planes have been rotated to lie in the common plane of the figure, so that the direction of the ideal tangential movement $du^{pq,\text{ideal},t}$ is to the right (positive ξ_1) and in the direction of the horizontal arrow. A bias (shift) is noted in this direction, with an average value $\xi_1 = 1$, which corresponds to the ideal movement $d\mathbf{u}^{pq,\text{ideal},t}$ projected onto the tangential plane. The particle arrangement in this assembly was initially isotropic (at zero strain), and the peak strength was reached at the much greater compressive strain of about 10%. If all particles moved in accord with an affine field, each dot would have the values $\xi_1 = 1$ and $\xi_2 = 0$ and all dots would appear as a single, shared point on the horizontal axis. Yet even at the very small strain of $\varepsilon = 0.0001$ in Figure 4.6, the relative tangential particle movements differ greatly from the ideal, affine movement. Some particles move at rates many times faster than those of the affine field and in directions that depart from the ideal direction (the positive ξ_1 direction in the figure), such that many of the relative movements between particle pairs are in the opposite, "wrong" direction. Clearly, the motions of individual particles are far more varied (and vigorous) than those predicted by the affine conditions of equation [4.5]. The density plots on the top and right of Figure 4.6 show the distribution of contact motions in the aligned ξ_1 and transverse ξ_2 directions, demonstrating quite large non-affine fluctuations, even near the start of loading.

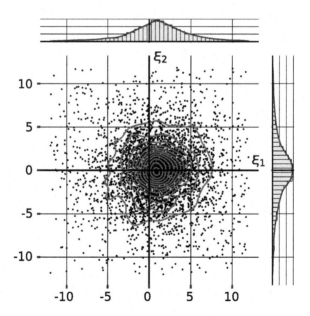

Figure 4.6. *Relative tangential movements $d\mathbf{u}^{pq,t}$ of particle centers during quasi-static triaxial compression of a DEM assembly of spheres. The value $\xi_1 = 1$ corresponds to perfect conformance of the actual movement $d\mathbf{u}^{pq,t}$ to the ideal tangential movement $d\mathbf{u}^{pq,\text{ideal},t}$*

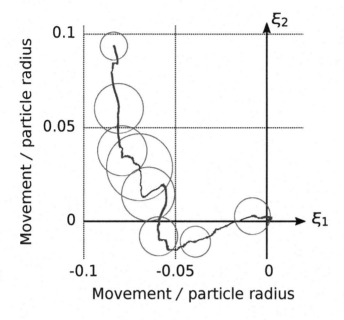

Figure 4.7. *Migration of a contact across the surface of a particle during triaxial compression of an assembly of spheres*

Figure 4.7 is another illustration of the discrepancy between ideal and actual movements. This figure is taken from a DEM simulation of the triaxial compression of a large assembly of spheres (see [KUH 16a]). The figure shows the movement of a single contact between two spheres, one of about 8500 contacts within the assembly. The two particles touched when loading was at the critical state and remained in contact for a bulk strain interval of about 0.8%, after which they separated. The plane of the figure represents a small portion of the surface of the first of the two particles (say, particle "p") as the second particle "q" moves across p. The movement of the contact point across p is computed with equation [2.10]. The small portion of the surface of p has been rotated so that the direction of the ideal affine tangential movement, $d\mathbf{u}^{c,\text{ideal,t}}$, is to the right (positive ξ_1). The contact in Figure 4.7 is seen to wander in a seemingly erratic manner particle radius, or about 30 times greater with much of the motion in the direction *contrary* to that of the bulk deformation (i.e. contrary to the ideal, affine direction of positive ξ_1). The contact's movement has been normalized by dividing by the radius of the particle, so that the movement of this one contact can be compared with the bulk strain. If fully unfolded, the normalized length of this contact's path is

0.22 times the particle radius, or about 30 times greater than the elapsed strain of only 0.008. That is, when viewed at the micro-scale of contact interactions, the migration activity of this contact is much more vigorous than the bulk deformation.

The circles in Figure 4.7 show the size of the Hertz contact area (computed with equation [2.23]) at various stages of its migration. The contact area is seen to expand and contract as the contact migrates, corresponding to a growing and diminishing normal force. The extent of rolling is also apparent, as the contact path is much longer than the diameters of the circular contact areas between the particles (radius a in Figure 2.2). Migration of the contact area occurs during periods of slip (shown as thicker lines) and also while the contact is not undergoing slip (shown with thinner lines). Frictional slip repeatedly starts and stops, occurring in a sporadic, fitful sequence of about 20 slip episodes. No aspect of this contact's behavior is regular: force and movement progress in an irregular, desultory fashion with numerous periods of slow, lingering movement and of darting, rapid movement.

Although particle movements are highly variable and differ greatly from their ideal, affine movements, a number of trends in the movements (in both normal and tangential directions) and rotations can be distinguished by carefully analyzing the data from simulations and physical models. Duran *et al.* [DUR 10] conducted DEM simulations of the isotropic compression and triaxial compression of dense sphere assemblies. By periodically constructing a 3D Delaunay tessellation of the assembly (Figure 1.2(c)), they analyzed the relative movements of particle pairs $d\mathbf{u}^{pq}$ that formed the edges of the tessellation's tetrahedra. Some edges corresponded to contacting particles (i.e. actual contacts), whereas other edges were between nearby particles that were not touching (called *virtual contacts*). On average, the relative movement $d\mathbf{u}^{pq}$ of all edges was close to the ideal movement $d\mathbf{u}^{pq,\text{ideal}}$. This average conformance is applied to both the normal and the tangential components of the relative movements across the edges. The conformance was quite different, however, for the actual contacts and the virtual contacts. Among the actual contacts, the average normal movements $du^{pq,\text{n}}$ were less than the average ideal normal movements $du^{pq,\text{ideal,n}}$ during both isotropic compression and triaxial compression. The normal movements of the actual contacts were about 10% less during isotropic compression. With triaxial compression at small strains, the average normal movements of the actual contacts were 30% less than the ideal normal movements; but at large strains (at the peak strength and beyond), the average actual normal movements were nearly zero, far less than the ideal movement. Duran *et al.* attributed these small normal movements to the high inter-particle stiffness, which limits the relative displacements of contacting particles in the normal direction. Contrarily, the normal movements of the virtual contacts exceeded the ideal normal movements. The opposite trend was noted for the tangential movements between pairs of particles. For the actual contacts during triaxial compression, the average tangential movements were greater than the ideal tangential movement (by about 5% at small strains and by about 50% at large strains). The larger average tangential movements of the actual contacts were compensated by smaller average movements among the virtual contacts.

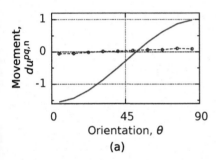
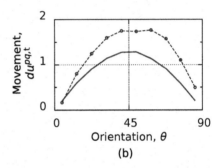

Figure 4.8. *Simulation results of biaxial compression of a disk assembly. Average relative motions of contacting particle pairs at various orientations θ relative to the direction of compressive loading: (a) normal movements $du^{pq,n}$; (b) tangential movements $du^{pq,t}$. Solid lines show ideal affine movements; dashed lines show the actual, measured movements [KUH 03a]*

A similar trend has been noted in 2D simulations of vertical biaxial compression of disk assemblies [KUH 03a]. Figure 4.8 shows the average incremental normal and tangential movements, $du^{pq,n}$ and $du^{pq,t}$, of contacting particle pairs at various orientations. The movements were computed at a compressive strain of 0.5%, which was near the peak state. At this stage of loading, the assembly expands in the transverse horizontal direction, and the two strain rates, $\dot{\varepsilon}_{11}$ and $\dot{\varepsilon}_{22}$, can be used in equation [4.4] to compute the ideal movements $du^{pq,\text{ideal},n}$ and $du^{pq,\text{ideal},t}$, which are also shown in the figure. The horizontal axis in Figure 4.8 gives the orientation θ of branch vectors l^{pq} relative to the direction of compressive loading. All movements have been normalized by dividing by the mean particle diameter and by the compressive strain increment $d\varepsilon_{11}$. The average actual movements in the normal direction are much smaller than, less than 10% of, those predicted with equation [4.4], but the average tangential movements are about 50% larger than the ideal movements.

In another simulation of biaxial compression of a disk assembly, the author compared the measured stress increments with the stress increments that would have resulted if particle movements were in perfect conformity with the ideal motions of equation [4.4], noting that the mean spin $\overline{\mathbf{W}}$ is zero for biaxial loading [KUH 04b]. The linear-frictional contact model in section 2.3.2 was used in the simulations and in the calculation of the ideal stress increments (the stiffness ratio α was 1). The results in Table 4.1 show the actual and idealized changes in the deviator stress $(dq = -(d\sigma_{11} - d\sigma_{22}))$ and pressure $(dp = -\frac{1}{2}(d\sigma_{11} + d\sigma_{22}))$ at the start of loading and at the peak and critical states. The changes in pressure resulted from increases in the compressive stress σ_{11} while maintaining a constant transverse stress σ_{22}. The results demonstrate that fluctuations in the particle movements (and rotations) from the ideal, affine condition have a significant effect on changes in stress. The effect of movement fluctuations is modest at the start of loading but becomes quite significant at large strains, where the fluctuations in movements and rotations conspire to produce a nearly stationary stress condition at the peak and critical states. Fluctuations are also seen to have a greater effect on the deviatoric stress than on the mean stress.

		Stress increments		
	$d\mathbf{u}^{pq}$	Initial loading	Peak state	Critical state
dq	Idealized	2610	4960	3500
	Actual	1760	0	0
dp	Idealized	1000	−275	235
	Actual	880	0	0

Table 4.1. *Stress increments during biaxial compression of disk assemblies: actual increments and those predicted with equation [4.4]. Increments in the deviatoric stress (dq) and pressure (dp) are in units of the contact stiffness k^n [KUH 04b]*

The conformity (or lack of conformity) between the actual and ideal relative movements, $d\mathbf{u}^{pq}$ and $d\mathbf{u}^{pq,\text{ideal}}$, can also be measured for distant non-touching particles. To this end, we consider distant pairs of particles that are part of the particle graph (Figure 1.3) and, thus, form the load-bearing framework of the material. A discrete metric for the distance between two such particles is shown in Figure 4.9(a), in which the distance between the reference particle (labeled "0") and another particle is the least number of contacts (graph edges) between the two particles. (These distances correspond to the $\mathbf{A}^{\text{dist}}$ matrix described near equation [4.27].) For example, particles (nodes) that are in direct contact with the reference particle have discrete distance 1, whereas particles that are twice removed have distance 2. The conformity of a pairs movements to the affine field can be quantified with the dimensionless inner product χ^{pq}:

$$\chi^{pq} = d\mathbf{u}^{pq} \cdot d\mathbf{u}^{pq,\text{ideal}} / |d\mathbf{u}^{pq,\text{ideal}}|^2 \qquad [4.10]$$

Particle movements in perfect conformity with the mean-field deformation have $\chi^{pq} = 1$; movements in counter-conformity have $\chi^{pq} = -1$; movements that are orthogonal to the mean-field movement have $\chi = 0$.

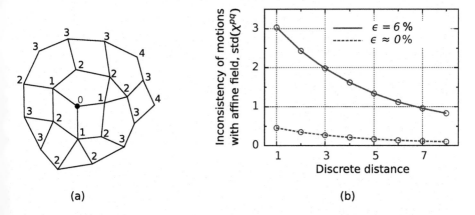

(a)   (b)

Figure 4.9. *(a) Discrete distances from a single particle (node) "•" to the surrounding particles, measured by the number of contacts that separate the particles. (b) Inconsistency of contact motions with the mean-field deformation during simulated triaxial compression of an assembly of sphere-clusters, expressed with the standard deviation of fluctuations from the mean field*

Figure 4.9(b) gives the dispersion of χ-values measured in simulations of triaxial compression of an assembly of sphere-clusters at two strains [KUH 16a]. The results are given as the standard deviations of χ at the start of loading and at the peak state (at compressive strain $\varepsilon = 6\%$) for assemblies of sphere-clusters undergoing triaxial compression. To place these results in perspective, the *mean* values of χ were very close to 1.0 at both strains and at all distances. That is, *on average*, the motions of particle pairs were consistent with the mean deformation. The average inconsistency (dispersion or fluctuation) in conformance is indicated by the standard deviation of χ, measured at different distances (Figure 4.10(b)). The figure shows that conformance improves with the distance (discrete separation) between particles. Moreover, particle movements conform more closely to the mean-field deformation at the start of loading $\varepsilon = 0$ than at the larger strain. Even at the start of loading, however, conformance is far from perfect, with the standard deviation std(χ) equal to 0.46 for contacting particles and 0.20 for particles that are separated by four contacts (at the ends of chains of five particles). Again, these standard deviations can be compared with the mean of 1.0, demonstrating large deviations of the particle movements from the affine field, even for distant particles.

Such results confirm a significant disparity between particle movements and the ideal movements of an affine field. Fluctuations in the motions are large and seemingly erratic, particularly during deviatoric loading near and beyond the peak strength. Under some conditions, the average disparity of the inter-particle movements from the affine movements is more modest: in particular, during isotropic compression and during deviatoric loading at small strains. Inter-particle movements at large deviatoric strains occur primarily in the tangential direction, whereas particle movements normal to the contact plane are largely suppressed. As demonstrated in Table 4.1, fluctuations of movements relative to those of the affine field greatly reduce the incremental stiffness of granular materials.

Notwithstanding the large variability in particle movements, some general trends can be noted, even at large deviatoric strains. The general trend of tangential movement is in the direction given by equation [4.7$_2$], although, on average, tangential movement proceeds at a somewhat faster rate. This general trend is shown in Figure 4.10, which gives the results of DEM simulations of plane-strain biaxial compression of an assembly of spheres at the critical state [KUH 10]. The arrows in this figure are superposed on a unit sphere. Each arrow shows the average directions and magnitudes of the tangential motions of those particle pairs having the general orientation given by the arrow's location on the unit sphere. That is, the unit sphere represents the orientations of the contact normals $\mathbf{n}^{pq}$ of contacting particles, and arrows represent the average tangential movements $d\mathbf{u}^{pq,t}$ of those particles having these orientations. In general, particles tend to migrate in a tangential manner away from the direction of bulk compression (the e_1 direction in Figure 4.10(a)) and toward the direction of bulk extension (direction e_3). These directions are the eigen-directions of the average rate of deformation tensor $\overline{D}$ in equations [1.94$_1$] and [4.6]. The plane-strain e_2 direction of these simulations is neither a source nor sink for the averaged tangential movements of particles. That is, the directions of the arrows in Figure 4.10, which represent averaged tangential movements, are nearly aligned with the ideal movements of equation [4.7$_2$]. At small strains, the average magnitudes of $du^{pq,t}$ and $du^{pq,\text{ideal},t}$ are also nearly equal. At the critical state, the averaged actual tangential movements are somewhat larger than $du^{pq,\text{ideal},t}$ by a factor of about 1.7 (see tangential movements in Figure 4.8(b)).

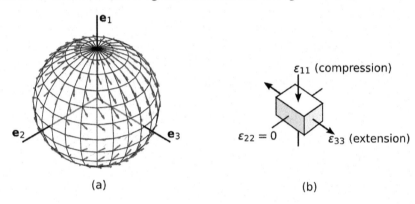

(a) (b)

Figure 4.10. *Unit sphere showing the general directions of tangential particle movements in DEM simulations of plane-strain biaxial compression of sphere assemblies. The assembly was compressed in the e_1 direction and extended in the e_3 direction, while maintaining plane-strain conditions in direction e_2. Arrows show the average direction and magnitude of tangential movements $du^{pq,t}$ for those contacts having various orientations [KUH 10]*

4.2.3. *Local strains*

As would be expected from the large deviations of the movements du^{pq} for neighboring and distant particle pairs from the ideal movements $du^{pq,\text{ideal}}$, the local strains, when measured at the meso-scale of particle clusters, are also highly heterogeneous and vary greatly from the bulk strain [ALS 06]. Because the bulk stiffness of a granular material is usually much smaller than the stiffnesses of individual grains, bulk deformation is largely the result of deformation within the void space. Computing strain within the voids is particularly straightforward for two-dimensional assemblies, by partitioning the assembly into void polygons (Figure 1.3) and using the methods described in section 1.4.1.2 to compute the strains within these polygons. At the scale of these individual void polygons, deformation is highly heterogeneous, and this heterogeneity is expressed in the statistical dispersion of void strains and in the spatial patterning of this heterogeneity, the latter being the subject of section 4.7. Heterogeneity increases with increasing bulk strain, and at the peak stress, the strains within many void polygons are *counter*-aligned with the bulk strain, with as much as 20% of the void space making a *negative* contribution to the bulk strain [KUH 03a]. By analyzing deformations within voids, Kruyt and Antony [KRU 07] demonstrated that the deformation of disk assemblies at the peak stress and at larger strains is largely determined by the tangential motions of particles rather than by their normal movements. This dominant influence of tangential movements on strain is consistent with the trends in the movement that were noted above. Other studies have shown that the dilation induced by the deviatoric loading of dense disk assemblies can be attributed to those voids which are elongated and whose longer axes are oriented in a direction parallel or nearly parallel to the direction of the major principal compressive stress: the shortening of these voids in the direction of compression tends to "pry" them open, inducing a local dilation [KUH 99, NGU 09]. Voids with an elongation that is oriented in the direction of bulk extension tend to compress and reduce the bulk volume, as they are flattened by the applied compression.

4.2.4. *Particle rotations*

Much attention has been given to the rotations of particles that occur during loading and deformation. In the absence of a mean bulk spin $\overline{\mathbf{W}}$, particle rotations should ideally be zero (equation [4.4$_2$]), but simulations and physical experiments have shown that particles rotate vigorously during deviatoric loading and that rates of rotation can be much larger than the norm of the bulk strain rate, $|\overline{\mathbf{D}}|$, particularly for disk and sphere assemblies. Figure 4.11 shows probability densities of the rotation rates of disks, spheres and sphere-clusters (Figure 3.1(f)) during simulations in which the loading was symmetrically applied, such that $\overline{\mathbf{W}} = 0$ (see also [BAR 94, CAL 97]). The rates, which have been normalized by dividing by the compressive strain rate $\dot{\varepsilon}$, are quite large and varied, with rotation rates much larger than the rate of bulk deformation. These simulations show that particle rotations are larger for spheres than for disks, probably because the extra dimension affords greater freedom of movement. The simulations also show that particle rotations are much larger for disks and spheres than for other shapes (also see [NOU 03]).

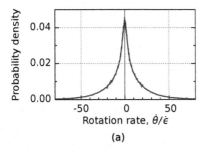 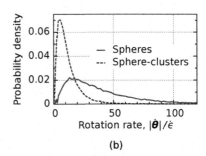

(a)   (b)

Figure 4.11. *Probability densities of particle rotations in loading simulations at the critical state: (a) disk assembly under biaxial loading; (b) sphere and sphere-cluster assemblies under triaxial loading*

In biaxial compression experiments on laboratory assemblies of aluminum, plastic and wood rollers, Misra [MIS 98] and Calvetti *et al.* [CAL 97] measured the rotations of individual rollers and found that the standard deviation of these rotations increased with increasing compressive strain and attained a value of about $12°$ at a compressive strain of 10%, evidence of rotation rates that were many times greater than the strain rate. Alshibli and Alranahi [ALS 06] measured the rotations of near-spherical pearl particles during triaxial compression and found particles that had rotated $60°$ when the global strain had only reached 23.5%. Although rotation rates are smaller for other particle shapes, the average magnitude of the rotation rate is typically several times the rate of bulk deformation, even with elongated polygonal particles [NOU 03, SAL 09].

Particle rotations tend to reduce the strength and stiffness of granular materials, and when they are suppressed by particle shape, by resistive moments at the contacts or by proscribing rotations with the simulation algorithm, strength is increased [BAR 92a]. Table 4.1 highlighted the large differences in actual stress increments and the increments that would have occurred if the particles had been constrained to move in accord with an affine field. Referring to this table, more than half of the difference between the constrained and actual increments of deviator stress dq can be attributed to the particle rotations [KUH 91, KUH 04b].

The effect of particle rotations on stiffness and strength can be viewed in the context of the relative movements $d\mathbf{u}^{c,\text{rigid}}$ between particles at their contacts (equation [2.2]). These relative movements produce incremental changes in the contact forces and in the aggregate, bulk stress. These relative movements can be separated into contributions from the particle translations and from the particle rotations:

$$d\mathbf{u}^{c,\text{rigid}} = d\mathbf{u}^{c,\text{trans}} + d\mathbf{u}^{c,\text{rot}} \qquad [4.11]$$

$$d\mathbf{u}^{c,\text{rigid}} = d\mathbf{u}^q - d\mathbf{u}^p \quad \text{and} \quad d\mathbf{u}^{c,\text{rot}} = d\theta^q \times \mathbf{r}^{c,q} - d\theta^p \times \mathbf{r}^{c,p} \qquad [4.12]$$

Although the rotations $d\theta^p$ and $d\theta^q$ can either increase or reduce the relative movement $d\mathbf{u}^{c,\text{rigid}}$, simulations with sphere and ellipsoid particles have shown that the dominant tendency is for rotations to reduce these relative movements, thus reducing bulk stiffness and strength [KUH 04d]. This conclusion is based on correlations between the translational and rotational contributions, $d\mathbf{u}^{c,\text{trans}}$ and $d\mathbf{u}^{c,\text{rot}}$, and between each of these contributions and the combined movement $d\mathbf{u}^{c,\text{rigid}}$. The translational and rotational contributions are negatively correlated and tend to counteract each other and reduce the movement $d\mathbf{u}^{c,\text{rigid}}$ (the Pearson correlation coefficient is between -0.2 and -0.65). The counteracting effect of the rotations is largest at large strains and also exists at small strains, but the particle rotations reduce both normal and tangential movements at the contact – both $d\mathbf{u}^{c,\text{rigid,n}}$ and $d\mathbf{u}^{c,\text{rigid,t}}$ in equation [2.3].

At small strains, the contact movements $d\mathbf{u}^{c,\text{rigid}}$ are closely correlated with the translational contributions $d\mathbf{u}^{c,\text{trans}}$, with coefficients of correlation greater than 0.8, but contact movements are entirely uncorrelated with the rotational contributions $d\mathbf{u}^{c,\text{rot}}$. Even though rotational contributions are uncorrelated with contact movements at small strains, the fluctuations in rotations are substantial, and they reduce the bulk material stiffness (as seen in Table 4.1). The opposite situation applies at large strains: the contact movements $d\mathbf{u}^{c,\text{rigid}}$ correlate weakly with the translational contributions $d\mathbf{u}^{c,\text{trans}}$ but correlate strongly with $d\mathbf{u}^{c,\text{rot}}$. That is, particle rotations assume a dominant role in contact deformation at large strains.

Although particle rotations are large and greatly influence strength and stiffness, most evidence suggests that the mean particle rotation is quite close to the mean bulk spin $\overline{\mathbf{W}}$ within a granular assembly. In a micro-polar setting, separate spatial fields are postulated for the mean spin of particles (the micro-spin $\overline{\mathbf{W}}^{micro}$) and the mean spin of the region within which the particles are embedded (the background spin $\overline{\mathbf{W}}$), as in equation [4.9]. Simulations of biaxial and triaxial compression (conditions in which the mean bulk spin is zero) of disks, ovals, spheres, and prolate and oblate ovoids demonstrate that the mean rotation of the particles is nearly zero [KUH 99, MAT 03, KUH 04d]. Simulations of the shear bands that develop during simple shear of sphere assemblies also show that the average particle rotations are equal to the local background spin throughout the thickness of a band [SUN 13]. The author conducted a series of 2D simulations of disk assemblies, in which a non-uniform deformation field was induced by applying body forces to the particles within an assembly. Although large shearing strains and meso-scale spins $\overline{\mathbf{W}}$ were coerced in this manner, the spatial distribution of mean particle rotations was nearly equal to the spatial distribution of mean spin, disaffirming the existence of a separate field of micro-rotation [KUH 05a]. Matsushima *et al.* [MAT 02] used laser-aided tomography (LAT) to track the motions and rotations of crushed glass particles in a model of slope failure and found that the grain rotations were about the same or a little larger than the underlying continuum rotation. Their simulations of shear band development in disk assemblies yielded grain rotations that were roughly identical to the continuum rotation. During their biaxial compression and simple shear tests with wood rollers, Calvetti *et al.* [CAL 97] found that the mean particle rotation was nearly equal to the mean rotation of the assembly. However, during tests of combined compression and shearing, in which the principal axes of strain were gradually rotated, they noted a drifting apart of the mean particle rotation and the specimen rigid rotation. Bardet and Proubet [BAR 91], in their study of shear band evolution in 2D simulations of disk assemblies, also measured a difference in the mean rotation of particles and the rotation of the underlying displacement field. In summary, although some studies have shown differences in the fields $\overline{\mathbf{W}}$ and $\overline{\mathbf{W}}^{micro}$, most have found that the two fields coincide. Any small differences in these fields are likely statistically (and mechanically) insignificant, given the large dispersion of the particle rotations, as shown in Figure 4.11.

4.2.5. *Contact rolling*

As discussed above, particle rotations play a role in the relative contact movements $d\mathbf{u}^{c,\mathrm{rigid}}$ (equations [4.11–4.12]). Particle rotations are also expressed in the rolling, twirling, twisting and rigid-rotations that arise in the interactions of particle pairs (section 2.1). The magnitudes of the rolling and rigid-rotation movements increase greatly with increasing strain and are greater for disks and spheres than for other shapes [SHO 03, KUH 04d]. Oda *et al.* [ODA 82] noted that the strength of granular materials is insensitive to the coefficient of friction between the grains (see section 4.6.3), and they attributed this observation to the effects of inter-particle rolling. Their experiments on lubricated and unlubricated ovals showed greater rolling activity among the unlubricated particles (also [SHO 03]). Experiments and simulations consistently demonstrate that rolling is the dominant mode of interaction among many more particles than sliding. A common misunderstanding, however, is that rolling reduces the tangential relative (sliding) movements $d\mathbf{u}^{c,\mathrm{rigid},t}$ between particles. We must make a distinction between particle rotations (the $d\theta^p$ and $d\theta^p$ in equation [2.2]) and contact rolling (defined as $d\mathbf{u}^{c,\mathrm{roll}}$ in equations [2.13] and [2.16] or as $d\theta^{c,\mathrm{rigid}}$ in equation [2.18]). Although particle rotations can reduce the relative tangential movement $d\mathbf{u}^{c,\mathrm{rigid},t}$ between particles (as discussed in relation to equations [4.11–4.12]), rolling and relative movement, $d\mathbf{u}^{c,\mathrm{roll}}$ and $d\mathbf{u}^{c,\mathrm{rigid},t}$, are distinctly different types of movements: rolling is an objective average movement of two particles at a contact, whereas $d\mathbf{u}^{c,\mathrm{rigid},t}$ is an objective relative movement. Indeed, the rolling movement $d\mathbf{u}^{c,\mathrm{roll}}$ and the relative movement $d\mathbf{u}^{c,\mathrm{rigid}}$ are linearly independent quantities, as was discussed with equation [2.13] and seen by comparing equations [2.5] and [2.6]. In the creep-friction model of rolling friction (section 2.4.2), the rolling between particles does affect their tangential force, but this effect is minimal for the relative amounts of rolling and sliding that typically occur in granular materials.

4.3. Coordination number, redundancy and contact longevity

The number of contacts among the particles in a granular material is a fundamental aspect of the material's state. Because bulk stiffness originates in the contacts' stiffnesses, a greater number of contacts will produce a greater bulk stiffness. A larger number of contacts produce more compelling restrictions on the particles' movements, reduce the rotations of particles, impart greater stiffness at the meso- and macro-scales and increase a material's dilation tendency. The average coordination number $\bar{n}$ in equation [1.15] is a gross measure of the number of contacts per particle. In naturally deposited sands, some particles can settle among their neighbors, creating contacts but without directly contributing to the load-bearing contact network. As noted in section 1.2.5, the zero-gravity environment of numerical simulations can produce a large fraction of rattler particles with no contacts and can also produce an additional fraction of inert, "bumper" particles with only one (2D) or two (3D) contacts, which do not contribute to the load-bearing network [THO 00]. The effective average coordination number $\bar{n}_{\text{eff}}$, which excludes these neutral, non-contributing particles and contacts, is:

$$\bar{n}_{\text{eff}} = 2\frac{M_{\text{eff}}}{N_{\text{eff}}} = 2\frac{M - M_{\text{bumper}}}{N - N_{\text{rattler}} - N_{\text{bumper}}} \qquad [4.13]$$

where N_{rattler}, N_{bumper} and M_{bumper} are the numbers of neutral particles and contacts. Because it is a measure of the relative number of contacts within the load-bearing particle network, Thornton [THO 00, GON 12] refers to $\bar{n}_{\text{eff}}$ as the *mechanical* coordination number. With non-convex particles, we must also distinguish between the number of neighboring (touching) particles and the number of contacts among these particle pairs, with possibly multiple contacts between neighboring pairs of particles. For these non-convex particles, the average number of contacts per particle is written as $\bar{n}^*$.

During loading under drained conditions, in which the lateral stress or the mean stress is maintained constant, the number of contacts (and the average coordination number) of medium and dense assemblies generally decreases during deviatoric loading, whereas the number of contacts in very loose assemblies increases [THO 00, ROT 04, GUO 13]. Both dense and loose assemblies eventually attain the same average coordination number at the critical state. These trends follow the trends in density noted in section 4.1. During drained loading, $\bar{n}_{\text{eff}}$ reaches a value somewhat greater than 3 for disk assemblies at the critical state and greater than 4 for sphere assemblies, values that are greater than the isostatic value described below. Rattler and bumper particles typically comprise 10–20% of all particles during zero-gravity drained loading simulations of disk and sphere assemblies [KUH 14a]. The trends in coordination number are more complex during undrained loading: the average coordination number of loose assemblies can decrease during loading, as contractive behavior causes the mean stress to decrease, producing a more fragile fabric with fewer contacts [GUO 13]. With dense assemblies in undrained loading, the dilation tendency, which is suppressed by an increasing mean stress, causes the coordination number to increase.

New contacts are constantly created during deviatoric loading, while existing contacts shift in their orientation and are separated. As discussed in section 4.3.2, the lifespan of a typical contact, as measured by the strain interval over which it exists, is quite brief, and the changing number of contacts (and the changing average coordination number) is the result of a difference in the rates of contact creation and separation. The rate of contact creation tends to be greater in directions of compression (i.e. negative normal strain or negative stretch), whereas the separation rate dominates in directions of extension [KUH 99, KRU 12]. The rate of change in the coordination number of both dense and loose assemblies is greatest at the start of deviatoric loading or at the start of load reversals during cyclic loading, as the initial contact network is disrupted by fresh loading or by a change in the loading direction [KUH 16b]. With initially dense assemblies in drained loading, the reduction in coordination number is primarily due to the loss of contacts in directions of bulk extension [CUN 82, KRU 12].

Other factors affect the coordination number during loading. The value of the average effective coordination number, $\bar{n}_{\mathrm{eff}}$, decreases with the increasing friction coefficient μ of the contacts, since large coefficients enable a larger number of particles to have only two or three contacts and a greater prevalence of zero-contact, rattler particles [THO 00, HUA 14, KUH 14a]. Particle shape also affects the average coordination number. The effective coordination number at the critical state is smallest for disks and spheres, and increases with decreasing sphericity and roundness [AZÉ 10, BOT 13, AZÉ 13b]. With non-convex particles, the number of contacts per particle, $\bar{n}^*$, increases with increasing non-convexity due to the increasing number of particle pairs having multiple contacts [AZÉ 13b].

4.3.1. *Redundancy*

The number of contacts is a scalar indicator of the stability of the particles' arrangement during loading. The *statical redundancy* of a granular region is a direct measure of the excess of contacts above the number required to maintain equilibrium and to maintain a stable load-bearing network of contacts. To evaluate redundancy, we compare the number of equilibrium constraints (and perhaps constitutive constraints) with the number of contact forces that are available to satisfy these constraints. In relation to the statics matrix of equation [3.79], the redundancy is a comparison of the numbers of its rows (i.e. the number of equilibrium equations) and its columns (i.e. the number of available contact force components). Because of the close relationship between the statics and kinematics matrices, an alternative *kinematic redundancy* is based on the number of kinematic (compatibility) constraints and the number of available particle movements within an assembly. The number of statical constraints and available forces is given in Table 4.2 for both two- and three-dimensional assemblies, for both convex and non-convex particles and for contacts both with and without contact moments. For example, the number of statical constraints in two dimensions is three times the number of particles: two force constraints and one moment constraint per particle. The "available forces" in the table include components of both contact force and contact moment. For example, two forces, normal and tangential, are associated with each contact in two-dimensional assemblies. When particles can transmit moments through their contacts, these additional moments are available to satisfy equilibrium and they increase the redundancy of the system.

	Number of constraints	Number of available forces
2D	$3N_{eff}$	
Convex particles, no contact moments		$2M_{eff}$
With contact moments		$3M_{eff}$
Non-convex contacts, no contact moments		$2M_{eff} + M_{2^+}$
3D	$6N_{eff}$	
Convex particles, no contact moments		$3M_{eff}$
With contact moments		$6M_{eff}$
Non-convex particles, not contact moments		$3M_{eff} + 2M_{2^+} + M_{3^+}$
Sliding contacts, friction limit	$(+)\,\eta M_{eff}$	

Table 4.2. *The number of statical constraints and available contact forces in a granular medium. The difference between forces and constraints is the absolute redundancy*

A pair of non-convex particles can touch at multiple contacts. This additional complexity is accounted for in the table: the number of neighboring particles with two contacts, two or more contacts, and three or more contacts are designated as M_2, M_{2+} and M_{3+}. In two-dimensional systems, two available force components (normal and tangential) are associated with each single contact pair of neighboring particles, but the multiple forces of a multi-contact pair can be resolved into a total of two forces and a contact moment. The table includes additional available forces for each of these particle pairs (a total of M_{2+} additional forces in two-dimensional systems). For non-convex three-dimensional particles, two additional forces are available for pairs with two contacts (which restrict a twisting and rocking of the pair), and pairs with three or more contacts have an additional available force.

The table also includes constraints on the forces that arise when some of the contacts are known to be sliding, as the normal and tangential forces must satisfy the friction limit (equality in equation [2.34]). Of course, we must not only know whether a particular contact is sliding but must also ascertain the direction of this sliding, before a constraint on the contact's force can be established. If η represents the fraction of contacts that are sliding, additional ηM_{eff} constraints are included in the ledger of Table 4.2. In this manner, each sliding contact is assumed to add a single extra constraint, since the magnitude of the tangential force is known to equal μf^n. Gong [GON 12] proposed two constraints for each sliding contact in three-dimensional systems, by assuming that the direction of sliding within the sliding plane is also known. In an alternative approach, the frictional limit is treated as a restriction on the available contact forces, and the term ηM_{eff} is subtracted from the number of available forces rather than added to the number of constraints [KRU 10, GON 12].

The ledger of available forces and constraints establishes a system's redundancy, with the absolute redundancy equal to the difference between the available forces and the equilibrium restraints. For the simple case of two-dimensional convex particles with no contact moments, the quotient of these two numbers is a measure γ_1 of relative redundancy:

$$\gamma_1^{2D,\text{convex}} = \frac{2\,\overline{n}_{\text{eff}}}{6 + \eta\,\overline{n}_{\text{eff}}} \qquad [4.14]$$

where $\bar{n}_{\text{eff}}$ is the effective average coordination number in equation [4.13]. Particle arrangements with a $\gamma_1^{2D,\text{convex}}$ greater than 1 indicate a redundancy of the contact network: the system is statically indeterminate, such that the space of available contact forces that fully satisfy the requirements of equilibrium and sliding has positive dimensions, and multiple combinations of contact forces can satisfy these requirements. Of course, the actual contact forces will also depend on the contacts' stiffnesses and on the fabric produced by the deposition and deformation history of the assembly. In Kruyt's [KRU 10] approach, sliding contacts reduce the numerator rather than augment the numerator, leading to a second measure of relative redundancy:

$$\gamma_2^{2D,\text{convex}} = \frac{\bar{n}_{\text{eff}}(2 - \eta)}{6} \qquad [4.15]$$

Both $\gamma_1^{2D,\text{convex}}$ and $\gamma_2^{2D,\text{convex}}$ are nearly equal when either is close to 1. A third measure neglects the influence of sliding contacts and is simply:

$$\gamma_3^{2D,\text{convex}} = \bar{n}_{\text{eff}}/3 \qquad [4.16]$$

Slowly deformed granular systems with a redundancy $\gamma_3^{2D,\text{convex}}$ greater than 1 are said to be over-constrained, statically indeterminate, hyperstatic or jammed. Systems with values less than 1 are hypostatic and are relatively fragile, collapsing from one state of marginal stability to another during deviatoric loading.

Each of these three redundancy quantities is a gross measure applied to an entire two-dimensional granular region. Sub-regions of larger and smaller redundancy (and stability) can exist within a granular region, and these sub-regions intermingle, emerge, grow and disappear throughout a deformation process [WAL 11]. The gross redundancy $\gamma_3^{2D,\text{convex}}$ is associated with the jamming limit, at which a sparse assembly that is being sheared will seize if the volume is slowly reduced until a sufficient number of contacts are formed, raising $\gamma_3^{2d,\text{convex}}$ to a value near 1 that defines the statically determinate or isostatic condition.

Corresponding definitions for three-dimensional convex particles with no contact moments are found with the complementary quantities in Table 4.2:

$$\gamma_1^{3D,\text{convex}} = \frac{3\,\bar{n}_{\text{eff}}}{12 + \eta\,\bar{n}_{\text{eff}}}, \qquad \gamma_2^{3D,\text{convex}} = \frac{\bar{n}_{\text{eff}}(3 - \eta)}{12}, \qquad \gamma_3^{3D,\text{convex}} = \bar{n}_{\text{eff}}/4 \quad [4.17]$$

Figure 4.12(a) shows the evolution of the deviatoric stress q/p and of various structural parameters for simulations of biaxial compression of a disk assembly at constant mean stress. The table in Figure 4.12(b) gives these parameters at the critical state for the disk assembly and for an assembly of spheres having a similar distribution of particle radii. Both assemblies had an inter-particle friction coefficient $\mu = 0.5$. The fraction of sliding contacts η during the deviatoric loading of disks is small: less than 16% of contacts are sliding at any stage of loading. With sphere assemblies, the fraction was somewhat larger, with about 31% sliding at the critical state. The largest fraction η is reached at small deviatoric strains prior to the peak deviator stress, and η remains at a slightly lower, constant fraction until the critical state is reached. In the simulations of Kruyt [KRU 10], the fraction η decreases with increasing friction coefficient μ. The data in Figure 4.12(b) show that about 6% of disks are either rattler or bumper particles, but 31% of spheres do not participate in bearing the applied stress. By accounting for the fraction of sliding contacts and rattler particles, the relative redundancy can also be tracked during loading. The simulations show that the relative redundancy of $\gamma_1^{2D,convex}$ starts at 1.4 and is reduced during loading, and becomes only slightly greater than 1 at the critical state. For the sphere assemblies, redundancy $\gamma_1^{3D,convex}$ is reduced to just 1.14 at the critical state (Figure 4.12(b)). Kruyt [KRU 10] showed that $\gamma_2^{2D,convex}$ is very nearly 1 for disks loaded to the critical state; and the sphere data of Thornton [THO 00] found that $\gamma_2^{3D,convex}$ is less than 1.15 at the critical state.

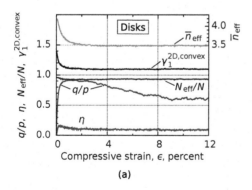

Conditions at the critical state

	Disks	Spheres
q/p	0.545	0.832
η	0.094	0.31
N_{eff}/N	0.94	0.68
$\bar{n}_{eff}$	3.49	5.16
$\gamma_1^{XD,convex}$	1.10	1.14

(a)   (b)

Figure 4.12. *Relative redundancy $\gamma_1^{2D,convex}$, fraction of sliding particles η, fraction of non-rattler particles N_{eff}/N and effective coordination number $\bar{n}_{eff}$: (a) values versus strain during biaxial loading of disk assemblies; (b) values at the critical state for assemblies of disks and spheres*

The average coordination number and relative redundancy are gross measures of the connectivity (topology) of granular materials. The number of contacts for individual particles will vary throughout an assembly, and data are best represented with a normalized histogram. Figure 4.13 shows histograms for an assembly of bi-disperse disks and an assembly of multi-disperse spheres taken during deformation at the critical state. Just as the average coordination numbers of initially loose and dense assemblies converge to a common stationary value at the critical state, the distribution of coordination numbers also converges to a characteristic distribution at the critical state [KUH 16b]. The critical state distribution depends on particle shape, the range of particle sizes and the friction coefficient μ between particles. Figure 4.13 shows that increasing friction shifts the distribution toward lower coordination numbers and greatly increases the number of particles with only two contacts and the number of rattler particles with none.

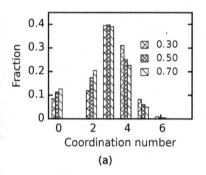 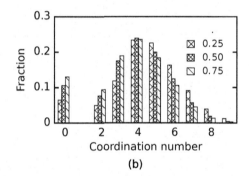

Figure 4.13. *Coordination number distributions at the critical state with various friction coefficients: (a) bi-disperse disks [KUH 14a] and (b) poly-disperse spheres [KUH 10].*

4.3.2. *Contact longevity and creation rates*

As has been noted, contacts are continually created and separated during deviatoric loading, and the average lifespan of a contact is usually quite brief. Kruyt [KRU 12] quantified contact lifespan as a half-life, defined as the deviatoric strain interval $\Delta\varepsilon$ at which half of the contacts that were present at strain ε are no longer present at strain $\varepsilon + \Delta\varepsilon$. In simulations of biaxial compression with dense and loose assemblies of disks, the half-life strain was less than 2% at small strains, and the half-life was somewhat more prolonged at larger strains, with a half-life of about 5% at the critical state [KRU 12]. The author [KUH 16a] measured the longevity of contacts in three-dimensional assemblies of spheres and of sphere-clusters during simulations of triaxial compression. The median longevity (half-life) of contacts within the sphere assembly was very brief. Half of the contacts that were initially present at the start of loading had separated after a deviatoric strain of only 0.68%, whereas half of the contacts that had formed between spheres during the subsequent loading had separated within an elapsed strain $\Delta\varepsilon_{11}$ of only 0.018%. Sphere-clusters are non-convex, and a pair of particles can have multiple contacts (Figure 3.1(f)). The median longevity of neighboring sphere-cluster particles was somewhat more extended than with sphere particles, but the median longevity of individual contacts between neighboring sphere-clusters was a deviator strain of only 0.02%. This data confirms the brief lifespan of most contacts, even during slow quasi-static (non-collisional) deformation.

Another measure of contact longevity is the rate at which contacts are created or separated. In the author's simulations of triaxial compression of sphere assemblies, the rate of contact creation was 6.8 contacts per particle per 1% of deviatoric strain [KUH 16a]. The rates of contact separation were of similar magnitude, again demonstrating the ephemeral nature of the contacts in a deforming granular material.

Contacts are continually created and separated at all orientations, although the net rate (creation rate minus separation rate) tends to be positive for contacts that are oriented in directions that are roughly aligned with the principal compressive strain: at contact directions $\mathbf{n}$ in which the stretch rate $\mathbf{n} \cdot (\overline{\mathbf{D}} \cdot \mathbf{n})$ is negative (tensor $\overline{\mathbf{D}}$ is the average rate of deformation, section 1.4). Contrarily, the rate of separation is usually greater than that of creation in directions with a positive stretch rate. These trends are shown in Figure 4.14, compiled from simulations of biaxial plane-strain compression of sphere assemblies [KUH 10]. The figure, corresponding to deformation at the critical state, shows the net rate of contact creation, which is a function of the three-dimensional space of contact orientations $\mathbf{n}$. Directions $\mathbf{e}_1$, $\mathbf{e}_3$ and $\mathbf{e}_2$ are those of the principal directions of deformation $\overline{\mathbf{D}}$: for these simulations, the three directions correspond to compression, extension and the neutral plane-strain condition, respectively. The net rate of creation is positive for orientations near $\mathbf{e}_1$, is negative for those near $\mathbf{e}_3$ and is zero in the plane-strain direction. At the critical state, the distribution of contact orientations (expressed, for example, with the fabric tensor $\overline{\mathbf{F}}$, equation [1.16]) attains a steady, stationary condition. Given the data in Figure 4.14, we might wonder why contacts would not continually accumulate in the compression direction ($\mathbf{e}_1$ in these simulations). The reason is that contacts tend to shift and migrate, thus changing their orientations $\mathbf{n}$, even during their brief lifespans. This migration, described in section 4.2, counteracts the net rate of contact creation, by causing existing contacts to migrate away from orientations of accumulation and toward orientations of depletion, thus maintaining constant fabric at the critical state [DID 01, KUH 10, RAD 12].

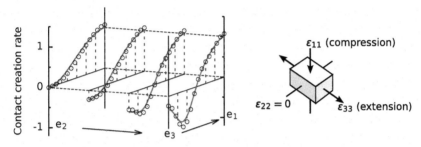

Figure 4.14. *Net rates of contact creation as a function of contact orientation, from simulations of biaxial plane-strain compression of sphere assemblies. The inset shows directions of compression (e_1), extension (e_3) and zero strain (e_2). Positive values indicate net contact creation; negative values indicate net contact separation [KUH 10]*

4.4. Anisotropy induced by loading

Loading of a granular material rearranges the particle orientations, the contact network and the distribution of forces within the network. These rearrangements give rise to an anisotropy of the material's internal structure and are largely responsible for its continued load-bearing capacity. Convincing evidence of this anisotropy was gained by Oda and his co-workers in the 1970s and 1980s with their careful inspection of resin-impregnated thin sections of sand specimens that had been extracted after triaxial compression loading [ODA 72a, ODA 72b]. This work revealed important anisotropies in the orientations of elongated particles and in the orientations of the contact normals. Extensive evidence of loading-induced anisotropy in the contact forces has also been obtained from photo-elastic experiments [DRE 72, ODA 85, MAJ 05] and from DEM and CD simulations. Among the most conspicuous manifestations of force anisotropy are the internal chains of heavily loaded particles (force chains) that are oriented roughly in the direction of the major principal compressive stress, as described in section 4.7.1.

4.4.1. *Anisotropy of particle orientations*

Several bulk measures of the orientations of the particle bodies and surfaces were introduced in section 1.2.4. Figure 4.15(b) shows the evolution of the particle-orientation tensor $\bar{\mathbf{J}}$ (equation [1.12]) in computer simulations of biaxial plane-strain compression of assemblies of elongated particles. The four simulations included dense assemblies of spheres and of smooth flattened (squat) ovoid particles, similar to oblate spheroids, having three different height-to-width ratios α: ratios 1.0 (spheres), 0.8, 0.625 and 0.50 "(the latter being the flattest, see Figure 3.1(c)). The particles in each assembly were initially arranged in an isotropic manner, with no preferred orientation of the particles [KUH 15], and changes in the deviator stress for the subsequent plane-strain loading are shown in Figure 4.15(a). The plot of the deviator of $\bar{\mathbf{J}}$ in Figure 4.15(b) shows that compression in the x_1 direction tends to rearrange the particles so that the flatter directions of the particles were preferentially oriented in the direction of compression, while the elongated direction became progressively oriented in the direction of extension (direction x_3). A stress-induced anisotropy is clearly produced from the initially isotropic arrangements of the non-sphere particles. Fairly large strains are required, however, to fully induce this anisotropy. With the assembly of the flattest ovoids (α = 0.5), a stationary fabric of particle orientation was not attained at even the largest strain of 60% – when the assembly was reduced to 40% of its original height – and with the ovoid assemblies, rearrangement continued to occur at strains well beyond the peak state, which was reached at a strain of only 5%. The results show that particle reorientation lags the change in stress during the early stage of loading, fabric change is a prolonged process and a critical state (steady state) of fabric is not necessarily attained when the volume and stress have become constant. Two-dimensional simulations of ellipse assemblies by Rothenburg and Bathurst [ROT 93] and three-dimensional simulations of prolate spheroids by Ouadfel and Rothenburg [OUA 01] show similar results: anisotropy in particle orientations continues to increase at large strain, even as the material approaches the critical state. (Another example of the lagging change in particle orientations is shown in Figure 4.23.)

Equation [1.14] defines a related bulk measure of particle orientation, based on the orientations of the particles' surfaces. In the author's simulations that produced Figure 4.15, the induced anisotropy of $\overline{S}$ that developed during deviatoric loading of sphere and ovoid assemblies is similar to that plotted for $\overline{J}$, demonstrating that the fabric tensors of the particle bodies and particle surfaces, $\overline{J}$ and $\overline{S}$, can both serve as measures of the anisotropy of particle orientation.

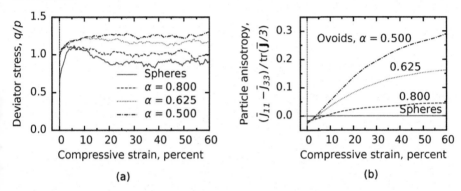

Figure 4.15. *Biaxial plane-strain compression in DEM simulations of assemblies of spheres and of flattened spheres (ovoids, Figure 3.1(c)): (a) stress and strain; (b) bulk anisotropy of particle orientations, using the measure $\overline{J}$ of equation [1.12] [KUH 15]. The flattest particles have aspect ratio $\alpha = 0.50$*

4.4.2. Anisotropy of contacts

As was seen in section 4.3, the average coordination number of initially dense assemblies is reduced during deviatoric loading, particularly during the early stage of loading (e.g. $\overline{n}_{\text{eff}}$ in Figure 4.12). Although contacts are continually gained and lost during deviatoric loading, the early net loss of contacts (and reduction of the average coordination number) occurs primarily as a loss of those contacts that are oriented in directions of extension (Figure 4.14 and [KUH 99, THO 00]). Not only does the average coordination number change, loading also alters the distribution of the numbers of contacts per particle. In the zero-gravity condition that is commonly used in DEM simulations, many particles lose all contacts and momentarily float or rattle among their neighbors (i.e. "rattler" particles, evident in values $N_{\text{eff}}/N < 1$ in Figures 4.12 and 4.13).

Three contact fabric tensors were introduced in section 1.2.5: the Satake tensor $\overline{\mathbf{G}}$ of contact orientation, the tensor $\overline{\mathbf{F}}$ of branch vector orientation and the mixed-orientation tensor $\overline{\mathbf{H}}$. The loading-induced evolution of the three fabric tensors is shown in Figure 4.16, for the simulations of plane-strain biaxial compression that were the subject of Figure 4.15. The figure shows the progression of the deviators of tensors $\overline{\mathbf{F}}$, $\overline{\mathbf{G}}$ and $\overline{\mathbf{H}}$ which have been normalized by dividing by their respective traces (conveniently, the traces of both $\overline{\mathbf{F}}$ and $\overline{\mathbf{G}}$ are 1). The increasing positive values in Figure 4.16(a) indicate that the contact normals become predominantly oriented in the direction of compressive loading (the x_1 direction in these simulations), with anisotropy increasing with advancing compressive strain. The more complex anisotropy of the branch vectors in Figure 4.16(b) is addressed below. The trend in the anisotropy of the contact normals during loading is a distinguishing characteristic of granular materials and has been widely reported [ODA 72b, ROT 89, OUA 01, GUO 13]. Figure 4.20(a) shows the stress-induced anisotropy of contact orientation in disk assemblies, with a rose diagram showing the orientations distribution that results from large deviatoric strains. The anisotropy that develops in the contact fabric is the basis of the stress–force–fabric relations that are discussed in section 4.4.4. Kruyt [KRU 12] and Oda *et al.* [ODA 85] have shown that the anisotropy in the Satake fabric $\overline{\mathbf{F}}$ at small strains is primarily the result of contacts being disengaged in the extension direction (also [ROT 92, KUH 99]). At larger strains, changes in $\overline{\mathbf{F}}$ are also produced by the re-orientations of existing contacts [KUH 10]. As seen in Figure 4.16, the anisotropy in contact orientation is larger for the less spherical shapes across the full range of strains [AZÉ 07]. For the sphere assemblies, this anisotropy reaches a peak value at 8–10% strain, which corresponds to the peak stress state (Figure 4.15(a)). At strains beyond 30%, the sphere assemblies have reached the critical state, in which stress, volume and contact fabric are stationary, a condition that is also seen in biaxial loading simulations of disks and spheres [PEÑ 07, KRU 12, ZHA 13b]. With non-spherical particles, more prolonged deformation is required to reach a steady fabric (see Figure 4.16(a)), further evidence of the significant fabric rearrangements that take place at large strains and an indication that a steady state of fabric is only attained at strains greater than 60%.

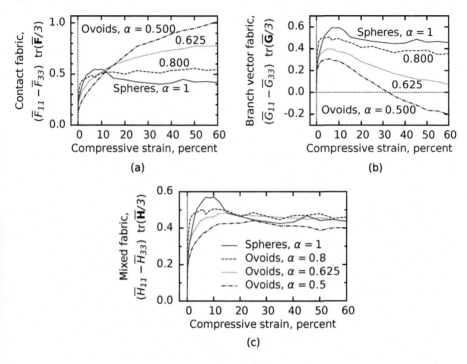

Figure 4.16. *Induced contact anisotropies in DEM simulations of plane-strain biaxial compression of assemblies of flattened spheres (ovoids): (a) contact normal vectors (equation [1.16]); (b) branch vectors (equation [1.18]); (c) mixed contact and branch vectors (equation [1.19]) [KUH 15]. The flattest particles have aspect ratio $\alpha = 0.50$*

The anisotropy of the branch vectors, as expressed with tensor $\overline{\mathbf{G}}$ in equation [1.18] and shown in Figure 4.16(b), is more complex. For spherical and nearly spherical particles, with $\alpha = 1$ and 0.8, evolution of the deviator $\overline{G}_{11} - \overline{G}_{33}$ follows a similar trend as the deviator of $\overline{\mathbf{F}}$ of the contact normals: branch vectors become increasingly oriented in the direction of compressive loading. The deviator of $\overline{\mathbf{G}}$, however, becomes negative for the flattest particles ($\alpha = 0.5$). For this particle shape, the contact normals become preferentially oriented in the x_1 direction of the compressive loading

(Figure 4.16(a)), but this type of anisotropy is accompanied by rotations of the flattened particles so that their wider dimension is preferentially oriented perpendicular to the direction of compressive loading (see Figure 4.15(b)). In the perpendicular direction x_3, the branch vectors tend to be longer and contribute to $l_3^c l_3^c$ in disproportion to their prevalence. Because the particle orientations require large strains to become fully expressed, the negative anisotropy of the branch vectors does not appear until strains of 30%. Of the three contact anisotropies, the evolution of the mixed anisotropy $\overline{\mathbf{H}}$ of equation [1.19] and Figure 4.16(c) most closely follows the progression of the deviator stress in Figure 4.15(a). As a further aspect of contact orientation, the orientation fabrics of so-called "weak" and "strong" contacts are discussed in section 4.5.

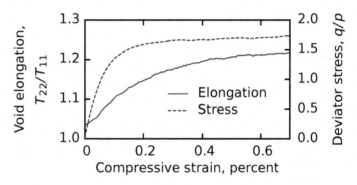

Figure 4.17. *Loading of disk assembly: deviator stress divided by mean stress and average ratio T_{22}/T_{11} of void widths, indicating elongation of the voids [KUH 99]*

4.4.3. *Voids anisotropy*

Several measures were presented in section 1.2.6 for characterizing the anisotropy of the voids space, including methods that are intended for either two- or three-dimensional materials. In two-dimensional assemblies, the average width of voids (i.e. the void polygons, as in Figure 1.3(a)) tends to be reduced in the direction of the principal compressive strain relative to voids' widths in the direction of extension. The average ratio in widths can be computed from the ratio of the eigenvalues of matrix $\overline{\mathbf{T}}$ in equation [1.24]. Figure 4.17 shows the evolution of this ratio during biaxial compression of a DEM assembly of disks, where T_{22}/T_{11} is the ratio of void widths transverse and aligned with the direction of compressive loading. During biaxial compression, the voids, on average, become progressively more elongated as they are flattened in the direction of compression [KUH 99, TSU 01, KRU 14], even as dilation causes a general increase in void size. Void elongation is most pronounced among those voids having larger valence [NGU 09]. An extreme display of such stress-induced anisotropy and void growth is found within thin localized shear zones (shear bands) where the post-peak deformation is often concentrated (section 4.7.4). Oda and Kazama [ODA 98a] presented evidence of large elongated voids within shear bands in their experiments with sands, physical disk assemblies and numerical simulations.

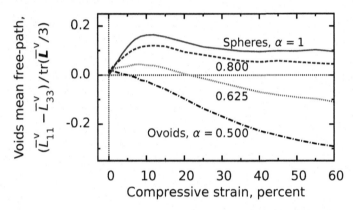

Figure 4.18. *Anisotropy of the median free path tensor* $\overline{\mathbf{L}}^{v}$ *of the voids during biaxial compression for assemblies of spheres and flattened ovoid particles (see Figure 4.15) [KUH 15]*

The evolution of void shape within three-dimensional assemblies of spheres and within those of flattened (ovoid) particles is shown in Figure 4.18, which gives the anisotropy of the median free path matrix $\overline{\mathbf{L}}^{v}$, defined in equation [1.42]. The figure is from DEM simulations of biaxial plane-strain loading and corresponds to Figure 4.15. The difference in median free path, $\overline{L}_{11}^{v} - \overline{L}_{33}^{v}$, is normalized by dividing by the trace of $\overline{\mathbf{L}}^{v}$. Because the assemblies dilate during loading, the trace progressively increases until the critical state is reached at strains of about 60%. Voids become more elongated in one direction (Figure 4.18) while they become narrower in the transverse directions. The spheres and the most rotund ovoids ($\alpha = 0.800$) develop voids that are longer in the direction x_1 of compression (positive values of

$\overline{L}_{11}^{v} - \overline{L}_{33}^{v}$), whereas the flatter ovoids develop voids that are longer in the extension direction. With the flatter particles, the void elongation continues to change at strains beyond 60%, providing further evidence that the fabric can change even after stress and density have attained steady conditions. The anisotropy of void orientation is consistent with an anisotropy of the material's permeability [SUN 13].

4.4.4. *Fabric anisotropy and stress*

Cundall [CUN 83] presented an elegant demonstration of the role of internal anisotropy in sustaining deviatoric stress. As with the other methods in this section, Cundall partitioned (and then reassembled) the stress within a granular assembly, thereby revealing its origin and nature. The expression for average stress in equation [1.53] can be partitioned into parts attributed to the normal and tangential contact forces. This partition is straightforward for disks and spheres, for which the contact vector $\mathbf{r}^{pq}$ between the center of particle p and its contact pq with particle q is aligned with the contact's outward normal vector $\mathbf{n}^{pq}$. The average stress $\overline{\sigma}$ in a region $\mathcal{B}$ of disks or spheres is:

$$\overline{\sigma}_{ij} = \overline{\sigma}_{ij}^{n} + \overline{\sigma}_{ij}^{t} \qquad [4.18]$$

$$= -\frac{1}{V^{\mathcal{B}}} \sum_{p \in \mathcal{B}} R^{p} \sum_{q \in Q^{p}} f^{n,pq} n_{i}^{pq} n_{j}^{pq} + \frac{1}{V^{\mathcal{B}}} \sum_{p \in \mathcal{B}} R^{p} \sum_{q \in Q^{p}} n_{i}^{pq} f_{j}^{t,pq} \qquad [4.19]$$

where $V^{\mathcal{B}}$ is the region's volume, R^{p} is the radius of p, Q^{p} is the set of particles that contact p, $f^{n,pq}$ is the compressive normal force and $\mathbf{f}^{t,pq}$ is the tangential force vector acting on p (Figures 1.4 and 1.6). Note that each contact is included twice in equation [4.18], once for each of its particles. Because the tangential force and the contact vector are orthogonal for disks and spheres, the stress contribution $\overline{\sigma}^{t}$ of the tangential forces is deviatoric (i.e. with zero trace).

Having accounted for the tangential contact forces, the remainder of the deviatoric stress is produced by anisotropies in the directions and distribution of the normal forces. This result was demonstrated by Cundall [CUN 83], again for disks and spheres, by partitioning the contribution $\overline{\sigma}^{n}$ in equation [4.18]:

$$\overline{\sigma}_{ij}^{n} = \overline{\sigma}_{ij}^{n,\,iso} + \overline{\sigma}_{ij}^{n,\,fabric} + \overline{\sigma}_{ij}^{n,\,var} \qquad [4.20]$$

$$= -\frac{1}{V^{\mathcal{B}}} \sum_{p \in \mathcal{B}} R^{p} \sum_{q \in Q^{p}} \frac{1}{d} \overline{f}^{n,p} \delta_{ij}$$

$$- \frac{1}{V^{\mathcal{B}}} \sum_{p \in \mathcal{B}} R^{p} \sum_{q \in Q^{p}} \overline{f}^{n,p} \left(n_{i}^{pq} n_{j}^{pq} - \frac{1}{d}\delta_{ij} \right) \qquad [4.21]$$

$$- \frac{1}{V^{\mathcal{B}}} \sum_{p \in \mathcal{B}} R^{p} \sum_{q \in Q^{p}} \left(f^{n,pq} - \overline{f}^{n,p} \right) n_{i}^{pq} n_{j}^{pq}$$

where $\overline{f}^{n,p}$ is the average of the compressive normal forces around p and d is the dimensionality (= δ_{ii}). For a particle p with m contacts, the average force $\overline{f}^{n,p}$ is $\frac{1}{m} \sum f^{n,pq}$. The first contribution in equation [4.21], $\overline{\sigma}^{n,\,iso}$, is isotropic and is equal to the volume-weighted average of the particles' mean stresses, since the inner sum of the first contribution is equal to a particle's volume times its mean stress, $V^{p}\sigma_{ii}^{p}/d$. The other two terms in equation [4.21] are deviatoric with zero trace. The second term, $\overline{\sigma}^{n,\,fabric}$, results from the anisotropy of contact orientations and is simply the deviatoric part of the Satake fabric tensor, weighted by particle size and particle mean stress (see $\overline{\mathbf{F}}$ in equation [1.16]). The final contribution, $\overline{\sigma}^{n,\,var}$, is due to anisotropy in the variation of normal contact forces: fluctuations from the average force that are correlated with contact orientation. Among the three contributions to deviatoric stress ($\overline{\sigma}^{t}$, $\overline{\sigma}^{n,fabric}$, and $\overline{\sigma}^{n,var}$), the anisotropies in contact orientation and in the directional distribution of normal forces each account for a greater portion of the deviatoric stress than the stress contribution of tangential contact forces, $\overline{\sigma}^{t}$ does.

Figure 4.19 shows the results of a DEM simulation of triaxial compression of a sphere assembly in which the assembly's height was reduced while maintaining constant mean stress. Among the three deviatoric contributions to stress ($\overline{\sigma}^{n,\,fabric}$, $\overline{\sigma}^{n,\,var}$ and $\overline{\sigma}^{t}$), the figure shows the greater importance of the anisotropies of the contact fabric and of the variations of the normal forces. These anisotropies of contact orientation and force are also apparent in the polar distributions of Figure 4.20. These results, similar to those of Rothenburg and Bathurst [ROT 89], are from DEM simulations of the vertical biaxial compression of multiple disk assemblies at the critical state. The three diagrams show the directional distributions of contact orientation, average normal force magnitude and average tangential force. A strong directional bias is seen in the contact orientations and averaged normal forces, both being larger in the vertical direction of compression, and with larger average tangential forces in oblique directions. Noting the scales of the plot axes, the relatively small values of the averaged tangential forces are due, in part, to the limitation imposed on these forces by the friction coefficient (for these simulations, $\mu = 0.5$), but the small values are also due to the partial cancellation of tangential forces having similar orientations but acting in opposite directions. That is, many contacts have a tangential force that acts in a direction countervailing to the deformation direction, and these forces partially cancel those tangential forces that act in the expected direction (see Figure 4.21(c) and the discussion on page 202).

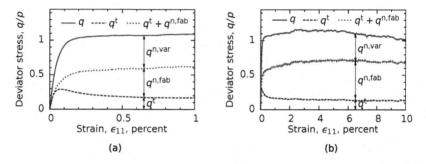

Figure 4.19. *Partition of the deviator stress from DEM simulations of triaxial compression of an assembly of 4096 spheres: (a) strain 0–1%; (b) strain 0–10%. The partition of equations [4.20]–[4.21] is applied to the deviator $q = \sigma_{11} - \sigma_{33}$ and normalized by dividing by the mean stress p*

Although equations [4.18]–[4.21] are an exact partition of stress for assemblies of disks or spheres, similar partitions for particles of general shape are more difficult to derive, but approximate stress–fabric–force (SFF) relations have been proposed for general shapes. These relations include the SFF approximations of Rothenburg and Bathurst [ROT 89] for ellipses, Ouadfel and Rothenburg [OUA 01] for ellipsoids, Azéma and Radjai [AZÉ 10] for pill-shaped ovals, Boton *et al.* [BOT 13] for smooth 3D platy particles and Azéma *et al.* [AZÉ 13b] for sphere-clusters. Their derivations begin with an assumption that the contact vector, contact force and contact normal vector can be treated as continuous vector-valued random variables, and a joint probability density $p(\mathbf{r}, \mathbf{f}, \mathbf{n})$ is assumed for the three quantities. In this manner, the sum in equation [1.51] is expressed as the integral:

$$\overline{\sigma}_{ij} = 2\frac{M}{V} \int_{\mathbb{R}^3} \int_{\mathbb{R}^3} \int_{\Omega} r_i f_j \, p(\mathbf{r}, \mathbf{f}, \mathbf{n}) \, d\Omega \, d\mathbf{r} \, d\mathbf{f} \qquad [4.22]$$

where $2M/V$ is the volume density of contacts (each contact being counted twice), Ω is the unit circle (2D) or unit sphere (3D) of orientations $\mathbf{n}$ of the contact normal and $\mathbb{R}^3$ is the full three-dimensional space of the possible contact and force vectors, $\mathbf{r}$ and $\mathbf{f}$ of Figures 1.4 and 1.7(b) (the former being limited by the sizes of the smallest and largest particles and the latter space being constrained by the constitutive character of the contacts). Noting that the joint probability $p(\mathbf{r}, \mathbf{f}, \mathbf{n})$ is the product of the conditional and marginal probabilities $p(\mathbf{r}, \mathbf{f}|\mathbf{n})$ and $p(\mathbf{n})$, equation [4.22] can be written as:

$$\overline{\sigma}_{ij} = 2\frac{M}{V} \int_{\mathbb{R}^3} \int_{\mathbb{R}^3} \int_{\Omega} r_i f_j \, p(\mathbf{r}, \mathbf{f}|\mathbf{n}) \, p(\mathbf{n}) \, d\mathbf{r} \, d\mathbf{f} \, d\Omega \qquad [4.23]$$

$$= 2\frac{M}{V} \int_{\Omega} \langle r_i f_j \, (\mathbf{n}) \rangle \, p(\mathbf{n}) \, d\Omega \qquad [4.24]$$

where $\langle r_i f_j \, (\mathbf{n}) \rangle$ is the expected value of the product $r_i f_j$ for those contacts with orientation $\mathbf{n}$.

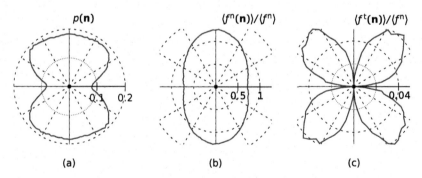

Figure 4.20. *Polar diagrams of directional distributions for the vertical biaxial compression of disk assemblies at the critical state: (a) density distribution of the contact normals; (b) distribution of the average magnitudes of the normal contact forces; (c) distribution of the average tangential forces*

Approximate stress–force–fabric relations are based on equation [4.24], with the quantity $Mp(\mathbf{n})/V$ representing the volume density of contacts having orientation $\mathbf{n}$. Density $p(\mathbf{n})$ can be approximated with a second- or higher-order Fourier series and expressed as a tensor function of $\mathbf{n}$ (as in equations [1.2–1.9]) or as a function of an orientation angle θ (as with equation [1.10] for two-dimensional systems). The contact distribution $p(\mathbf{n})$ in most stress–force–fabric relations is a second-order approximation, when applied to symmetric biaxial loading conditions (2D) or symmetric triaxial conditions (3D).

Approximating the average product $\langle r_i f_j (\mathbf{n})\rangle$ is more problematic, although it is usually simplified by assuming that the quantities $\mathbf{r}$ and $\mathbf{f}$ are uncorrelated (with zero covariance), so that their average product at orientation $\mathbf{n}$ is equal to the product of the averages of each quantity:

$$\langle r_i f_j (\mathbf{n})\rangle \approx \langle r_i(\mathbf{n})\rangle \langle f_j(\mathbf{n})\rangle \qquad [4.25]$$

where each bracketed quantity is a function of orientation $\mathbf{n}$. Contrary to this assumption, experiments and simulations have shown that the most prominent contact forces tend to act on larger particles, so that the approximation [4.25] leads to small errors in the stress–force–fabric relation. As a further approximation, the average contact vector $\langle \mathbf{r}(\mathbf{n})\rangle$ is sometimes assumed to have a magnitude that is independent of orientation, so that its components can be estimated as the average length $\langle |\mathbf{r}|\rangle$ multiplied by the average unit direction $\mathbf{e}(\mathbf{n})$ of $\mathbf{r}$ among those contact vectors having the normal direction $\mathbf{n}$, such that $\langle r_i(\mathbf{n})\rangle = \langle |\mathbf{r}|\rangle \langle e_i(\mathbf{n})\rangle$. The assumption of a common average length for all directions leads to errors of a few percent for disks and spheres, but the error can be quite significant for elongated particles [OUA 01, AZÉ 09, AZÉ 10].

The distribution of average force $\langle \mathbf{f}(\mathbf{n})\rangle$ among contact orientations $\mathbf{n}$ can be simplified by separating force into its normal and tangential components:

$$\langle f_j(\mathbf{n})\rangle = -n_j\langle f^{\mathrm{n}}(\mathbf{n})\rangle + \langle f_j^{\mathrm{t}}(\mathbf{n})\rangle \qquad [4.26]$$

The principle of self-equilibration requires that the scalar distribution $\langle f^{\mathrm{n}}(\mathbf{n})\rangle$ is an even function of $\mathbf{n}$, with $\langle f^{\mathrm{n}}(\mathbf{n})\rangle = \langle f^{\mathrm{n}}(-\mathbf{n})\rangle$, whereas the vector-valued distribution $\langle f_j^{\mathrm{t}}(\mathbf{n})\rangle$ must be an odd function of $\mathbf{n}$, with $\langle f_j^{\mathrm{t}}(\mathbf{n})\rangle = -\langle f_j^{\mathrm{t}}(-\mathbf{n})\rangle$. Noting that $\langle f^{\mathrm{n}}(\mathbf{n})\rangle$ is of the same form as the left side of equation [1.1], the distribution of normal force can be approximated as the mean normal force $\langle f^n \rangle$ multiplied by an n-th order probability distribution in the form of equation [1.2] or [1.10]. The odd-valued nature of tangential force, $\langle f_j^t(\mathbf{n}) \rangle$, complicates its approximation with the fabric tensors introduced in section 1.2.3. The situation is greatly simplified for two-dimensional assemblies that are loaded under symmetric biaxial conditions, for which we can adopt a convention of counterclockwise and clockwise tangential forces being positive and negative, respectively. With this convention, the distributions become even valued with an average value of zero across orientations of 0 to 2π. The vector-valued distribution $\langle f_j^t(\mathbf{n}) \rangle$ can then be approximated with the distributions of equations [1.2] and [1.10] by neglecting the leading constant 1 and using only the remaining, deviatoric tensors.

4.5. Density distribution of contact force

Each radial distance in Figures 4.20(b) and 4.20(c) represents an average magnitude of the normal or tangential force having a particular orientation of the contact normal. The contact forces among particles are not uniform, however, but have a wide dispersion of values. Typical probability densities of normal and tangential force magnitudes are shown in Figure 4.21. These distributions were extracted from simulations of bi-disperse disk assemblies undergoing biaxial compression at the critical state. The forces are normalized by dividing by the mean normal force magnitude $\langle f^n \rangle$ and plotted with semi-logarithmic axes. The distributions of normal forces are quite different for forces greater and smaller than the mean – for $f^n/\langle f^n \rangle > 1$ and $f^n/\langle f^n \rangle < 1$). The larger forces roughly conform to an exponential distribution, whereas forces smaller than the mean exhibit a flatter, "shouldered" distribution. Similar distributions of normal force have been reported for initial, isotropically compressed packings; for various stages of subsequent deviatoric loading; for a range of confining stresses; from contact dynamics (CD) simulations that enforce a rigid-contact constraint and from discrete element (DEM) simulations in which contacts have either linear or Hertzian stiffnesses; from laboratory experiments on assemblies of photo-elastic disks and from glass bead packings; for particles of different shapes and size distributions [RAD 96, THO 98, KRU 03, KUH 03a, AGN 07a, AZÉ 13b]. Although small variations in the distribution shapes from those in Figure 4.21(a) have been reported (a different slope or shape of the tail or a differently shaped distribution for the smallest forces), the primary trends in the figure are pervasive. Distributions of tangential forces similar to that in Figure 4.21(b) are also widely reported: a nearly exponential distribution for forces, sometimes with a reduced slope for the smaller forces [RAD 96, RAD 98, KRU 03, ZHA 10, AZÉ 13b].

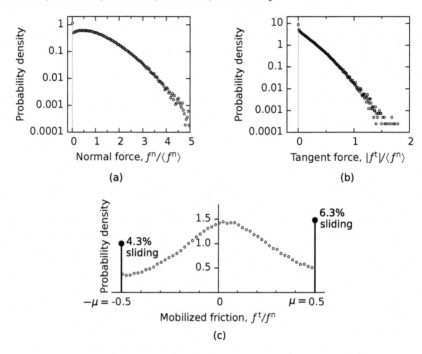

Figure 4.21. *Probability density distributions of contact forces in an assembly of bi-disperse disks loaded at the critical state: (a) normal contact forces; (b) tangential force magnitudes; (c) the ratio of tangential and normal forces. About 10.6% of contacts are sliding, with 6.3% sliding in the forward direction and 4.3% sliding in the reverse direction (see section 4.4)*

The shape of the normal force distribution might seem like an interesting but inconsequential characteristic of force transmission in granular materials, but Radjai and his co-workers [RAD 97, RAD 98] found that the macro-scale role of a contact is often determined by the magnitude of its force: "the inhomogeneous distribution of forces on the particle scale does not average out at the macroscopic level." The bimodal distribution of normal forces (a shouldered shape for small forces, but exponential for large forces) led these investigators to define two sets of contacts: *weak contacts* with normal forces less than the mean ($f^n/|f^n| < 1$) and *strong contacts* with normal forces larger than the mean ($f^n/|f^n| > 1$). During loading, the set of strong contacts is preferentially oriented in the direction of the major principal compressive stress, with a large fabric anisotropy and a fabric tensor that is aligned with the compressive stress tensor; but weak contacts are either isotropic in their distribution, slightly anisotropic in the direction of the major principal stress or have an anisotropy that is orthogonal to major principal stress [RAD 98, ZHA 11]. If the stress summation in equation [1.53] is calculated separately for the weak and strong contacts, most of the deviatoric stress (in most cases, more than 90% of the deviatoric stress) is attributed to the strong contacts [RAD 98, THO 98, EST 08, ZHA 11]. The deviatoric stress attributed to weak contacts can even be counter to the full deviatoric stress of the entire material, with a countervailing contribution to the full deviatoric stress. The weak contacts, however, primarily resist the mean stress: in simulations of sphere assemblies, Thornton and Antony [THO 98] noted that the weak contacts comprised about 70% of all contacts, but contributed 40% of the mean stress.

The different distributions of normal and tangential forces, as shown in Figures 4.21(a) and 4.21(b), also suggest that the mobilized friction f^t/f^n is larger, on average, among the weak contacts than among the strong contacts. Indeed, the likelihood that a contact will slide is negatively correlated with its normal force, most of the sliding contacts (sometimes more than 90%) are among the weak contacts.

For these reasons, Radjai *et al.* [RAD 98] proposed that the contact network be considered as two complementary sub-networks: a load-bearing network of strong contacts, which supports the deviatoric stress, and a dissipative network of weak contacts, which largely supports the mean stress. The sub-network of strong contacts includes most of the force chains that trend roughly parallel to the major principal compressive stress (section 4.7.1). The sub-network of weak contacts has been imagined to "prop" these force chains so that they can bear the bulk of the deviatoric stress [RAD 98, EST 08]. In section 1.2.5, strong fabric tensors $\overline{\mathbf{F}}^{\text{strong}}$, $\overline{\mathbf{G}}^{\text{strong}}$ and $\overline{\mathbf{H}}^{\text{strong}}$ were defined as complements of the assembly average tensors but are restricted to the strong contacts. The strong fabric tensor $\overline{\mathbf{H}}^{\text{strong}}$ has been found to correlate closely with the deviatoric stress tensor for spherical and non-spherical particles [ANT 04, THO 10, GUO 13]. The strong and weak contact anisotropies induced by loading are shown in Figure 4.22 for the simulations that were the subject of Figures 4.15 and 4.16. Comparing Figures 4.22(a) and Figure 4.16(c), the fabric of the strong contacts is very similar to that of all contacts, whereas the fabric of the weak contacts in Figure 4.22(b) is much less anisotropic than that of all contacts.

Figure 4.21(c) shows the density distribution of the mobilized ratios of tangential and normal contact force, f^t/f^n, for the biaxial loading of a disk assembly. A direction, positive or negative, is assigned to each tangential force $\mathbf{f}^t$ by comparing it with the direction that would be consistent with the bulk deformation rate $\overline{\mathbf{D}}$. Positive values correspond to tangential forces consistent with $\overline{\mathbf{D}}$, with a positive inner produce $\mathbf{f}^t \cdot d\mathbf{u}^{c,ideal,t}$, as defined in equations [4.6] and [4.8$_2$]. As shown in the figure, both consistent and countervailing tangential forces are present during deformation at the critical state. About as many contacts are undergoing "backward" slip as are slipping in the consistent, "forward" direction (4.3% versus 6.3%), and roughly 40% of the particles have a negative "backward" ratio f^t/f^n.

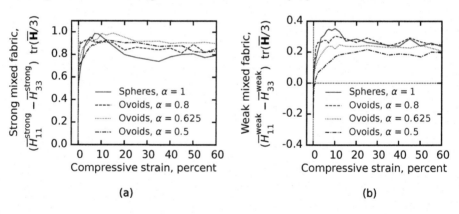

Figure 4.22. *Anisotropies in contact orientations among strong and weak contacts in DEM simulations of plane-strain biaxial compression of assemblies of flattened spheres (ovoids) using the anisotropy measure of equations [1.19] and [1.20]: (a) strong fabric of the mixed-tensor,* $\overline{\mathbf{H}}^{strong}$*; (b) weak fabric of the mixed-tensor,* $\overline{\mathbf{H}}^{weak}$

4.6. Effects of macro- and micro-scale characteristics

4.6.1. *Initial anisotropy*

An initial (or *inherent*) arrangement of particles is imparted by the processes of assembly, deposition, compaction and conditioning, and this initial arrangement affects the subsequent loading behavior (section 1.2). With elongated and flat particles, the directions of their longer axes are preferentially aligned perpendicular to the direction of deposition (gravity), whereas the directions of the contact normals are predominantly oriented in the deposition direction. The initial arrangement can be characterized with any of the scalar or tensor measures presented in section 1.2. Of the various fabric characteristics, the initial bulk density and the initial anisotropy of the particles and contacts have the greatest effect on stiffness, strength and volume change during loading (the effect of density was addressed in section 4.1), and experiments demonstrate that strength is largest and dilation is most intense when the direction of the major principal compressive stress is aligned with the original direction of deposition (gravity), and strength and dilation are smallest when the loading and deposition directions are perpendicular. Besides the experiments described in section 1.2, Oda and his co-workers [ODA 85, KON 82] conducted laboratory experiments with flat polyurethane ovals of two shapes that had been deposited (by hand) with an initial bedding direction and then rotated and loaded with biaxial compression in other directions. Consistent with tests on sands, they showed that strength and dilation were greater when the compressive loading was aligned with the original direction of deposition, and the effect of the loading direction was more pronounced with assemblies of more elongated ovals. These trends have also been reported for DEM simulations with assemblies of disks [CHE 90] and of convex polygons [HOS 12, NOU 05], in which particles were deposited under the influence of a virtual gravity. Regardless of any initial anisotropy, elongated particles tend to rotate during loading so that their longer axes become perpendicular to the direction of the principal compressive stress (section 4.4.1). If the directions of deposition and compressive loading are not aligned, the particles must undergo greater rotations during the loading process, requiring larger strain to reach the peak and critical states [NOU 05].

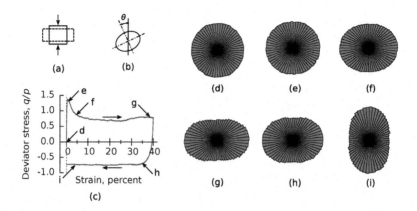

Figure 4.23. *Rose diagrams of particle orientations for elongated ellipse particles: (a) assembly subject to vertical compression, followed by vertical extension; (b) orientation of particles; (c) loading sequence; (d)–(i) distributions of particle orientation at strains 0%, 1%, 5%, 40%, 35% and 0%*

The rose diagrams of Figure 4.23 show the evolution of directions of particle elongation within an assembly of ellipse particles having a length-to-width ratio of 1.3. The initial fabric was isotropic, but a vertical shortening (loading) of the assembly produces a strong anisotropy of particle orientations. The sequence in Figure 4.23(d–g) shows that considerable deformation is required to fully reorient the particles during the initial loading phase. After reaching the condition shown in Figure 4.23(g), the loading was reversed, returning to the original zero vertical strain. An "initial anisotropy" applies to the start of this reversed loading, with an anisotropy that is perpendicular to the eventual, fully reverse-loaded condition. The unloading behavior is softer than that of the original loading and requires a reversal of over 10% percent strain (from 40% to 30%) to recover an isotropic distribution of particle orientations. An additional 30% of reversed loading (from strain 30% to 0%) is required to fully impart an anisotropic distribution of orientations. The predominant direction of particle elongation, which was initially horizontal, was rotated 90° during the course of reverse loading.

4.6.2. *Particle shape*

As was noted in section 1.1, grain shape is among the most important constitutional factors affecting the behavior of a granular material. Laboratory tests and numerical simulations of disks and spheres consistently yield strengths that are considerably lower than strengths measured in laboratory tests of natural sands and gravels and of simulations with particles of a more irregular shape. The small strength of disk and sphere assemblies is primarily due to the particles' greater propensity for rotation and rolling. Whereas the rotation of disks and spheres is resisted only by tangential contact forces, the normal forces between particles of other shapes also provide rotational resistance and contribute to the moment equilibrium of the particles (as occurs when $\mathbf{r}$ and $\mathbf{f}^n$ are not aligned in the $\mathbf{r} \times \mathbf{f}$ terms of equations [1.60] and [3.49]). In a compilation of the testing results of dozens of sands, Cho *et al.* [CHO 06] determined that the critical state friction angle decreases with increasing roundness and sphericity of the particles and was smallest for specimens of spherical glass beads. Shin and Santamarina [SHI 12] tested the angle of repose of mixtures of two sands – one with relatively rounded and the other with more angular shapes – and found that the angle of repose increases with an increasing fraction of angular particles. Stiffness, as measured with shear wave velocity, was also determined to increase with an increasing fraction of angular particles.

Numerical simulations have been used to examine the effects of several aspects of particle shape: elongation (lack of sphericity), non-convexity and angularity. In one of the earliest uses of the DEM to study the effect of elongated particle shapes, Rothenburg and Bathurst [ROT 93] simulated the biaxial compression of assemblies of ellipses having different elongations (aspect ratios). The peak friction angle increased with increasing particle elongation for ellipses for aspect ratios less than 1.35 but decreased slightly with larger aspect ratios. Contact dynamics (CD) simulations of two-dimensional elongated pill-shaped particles by Azéma and Radjai [AZÉ 10] showed a consistently increasing strength at the critical state with an increasing elongation of the particles, and CD simulations by Nouguier-Lehon [NOU 10] of assemblies of convex polygons having different elongations also demonstrated a consistent increase in strength at the critical state with increasing particle elongation. Similar to the results of Rothenburg and Bathurst [ROT 93], the peak strength increased with elongation until an elongation of about 1.4 but decreased slightly at larger elongations. The effect of non-convexity was also investigated in the simulations of Saint-Cyr *et al.* [SAI 12], who expressed the degree of non-convexity as a ratio of circumscribing and inscribing radii, similar to the sphericity definitions presented in section 1.1. For a variety of shapes, including disk clusters, ovals and polygons, strength at the critical state increased with increasing non-convexity.

Simulations of three-dimensional assemblies with different particle shapes show a less consistent effect of shape on strength. The author's DEM simulations of smooth oblate spheroids (flattened ovoids) gave a much larger strength (both at the peak and critical states) for the oblate shapes than for spheres [KUH 03b], but DEM simulations on prolate (lengthened, rugby ball) spheroids exhibited a smaller effect of the shape and a small *decrease* in strength with decreasing sphericity [OUA 01, NG 04]. CD simulations of other three-dimensional shapes have demonstrated a clearer influence of shape. The smooth plate-like shapes simulated by Azéma *et al.* (see Figure 3.1(e)) [BOT 13] gave a consistently greater strength with increasing flatness of the particles, and simulations of non-convex sphere-clusters showed a progressive increase in strength and a greater dilation tendency with increasing non-convexity [SAL 09, AZÉ 13b]. The effect of particle angularity was studied by Azéma *et al.* [AZÉ 13a] with CD simulations of convex irregular polyhedra, all having the same sphericity but with different angularities. Angularity was defined as the average angle between adjoining faces (π minus the dihedral angle, equal to 0 for a sphere, $\pi/2$ for a cube and somewhat larger, $\pi - \arccos \frac{1}{3}$, for a tetrahedron). Strength at the critical state increased consistently with increasing angularity.

4.6.3. *Inter-particle friction*

Experiments and simulations demonstrate that the bulk strength of granular materials increases with an increasing coefficient of inter-particle friction μ. Direct evidence was presented in the laboratory experiments of Konishi *et al.* [KON 82] on flat polyurethane oval particles, in which the peak friction angle ϕ was reduced from 52° to 26° by lubricating the particles with talcum powder (4.2). Simulations have shown that the peak and critical state friction angles increase greatly with μ for coefficients between 0.1 and 0.3, but that very little strength increase occurs for coefficients greater than 0.5 [OGE 98, THO 00, KRU 06]. That is, the dependence of strength on μ is nonlinear, and strength is relatively insensitive to μ for coefficients above 0.50 (see Figure 4.24).

An increase in μ produces a larger bulk small-strain stiffness, a greater dilation during loading and a higher void ratio at the critical state

[BAT 90, CHE 90, THO 00]. With larger μ, particle rolling is more vigorous, a smaller fraction of the contacts undergo slip during bulk deformation, and the sudden fluctuations in stress noted in section 4.1.1 are more frequent and intense. Fabric at the critical state also changes with an increase in μ: the number of contacts (and the average coordination number) is reduced [ZHU 95, THO 00, KUH 14a], the number of zero-contact rattler particles increases [HUA 14] and the anisotropy of contact orientations increases [CHE 90, HUA 14], with greater preponderance of contacts oriented in the direction of the principal compressive stress (i.e. greater anisotropy of the fabric tensor $\overline{F}_{ij}$ in equation [1.16]). The author has provided a rationale for the increase in coordination number with increasing μ for two-dimensional materials, as particles with only two contacts can remain stable for a wider range of geometries, thus increasing the prevalence of particles with low coordination numbers (see also [HUA 14]).

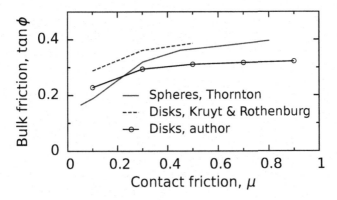

Figure 4.24. *Effect of friction coefficient μ on strength from simulations. Strength is expressed with the friction angle ϕ: triaxial compression of sphere assemblies by Thornton [THO 00], biaxial compression of disk assemblies by Kruyt and Rothenburg [KRU 06] and biaxial compression of disk assemblies by the author*

4.6.4. Other contact characteristics

Rotational, rolling resistance at particle contacts has been used in DEM and CD simulations to reduce the free rotation of particles (section 2.4 and [IWA 98, AI 11]). This computational device is professed to imitate the effects of the non-convexity and angularity of particle shapes and the natural capacity of such irregular shapes to reduce rotations. Under boundary conditions that inhibit the development of shear bands, the inclusion of a rotational friction-like resistance at the contacts has been shown to increase the peak bulk strength, increase bulk strength at the critical state, increase the dilation tendency, increase the void ratio at the critical state and increase contact anisotropy at the critical state, with these effects increasing with an increase in the rotational friction coefficient [ODA 97, EST 11, ZHA 13a, ZHO 13]. Indeed, Bardet [BAR 94] has shown that altogether preventing particle rotations more than doubles the strength of disk assemblies above that of freely rotating disks. Including rotational resistance at the contact has a minimal effect, however, on a material's modulus of elasticity and Poisson's ratio [MOH 10].

With boundary conditions that permit the localization of strain, a rotational friction promotes the formation of shear bands rather than a more diffuse, homogeneous failure within the material [IWA 98]. Rolling resistance also results in narrower shear bands, as the shear band thickness decreases with an increasing coefficient of rolling friction [MOH 10, ZHA 13a].

4.7. Localization and patterning

As was seen in section 4.2, particle movements typically depart significantly from the affine condition during quasi-static loading, and these non-affine, non-homogeneous movements are associated with a reduction in bulk stiffness and in the eventual failure of a granular material. Non-affine movements do not occur in a random, uncoordinated fashion, but are instead spatially organized and expressed in various patterns of localized deformation and force transmission. Several forms of such localization are discussed in this section. Localization occurs spontaneously within a granular material. Some forms of localization are expressed at the very start of loading, whereas other types do not appear until after extensive loading or at the onset of failure. Some forms, such as force chains and micro-bands, seem to be ubiquitous and natural, and are inevitable conditions of deviatoric loading. Some forms persist across a large range of strains, whereas other localization patterns are ephemeral, and arise and subside frequently during the loading process. Each type of localization feature discloses a coherent organization of movement and force which alludes to an internal length-scale that operates across distances the size of the feature. These internal length-scales should be contrasted with those of an ideal classical continuum, for which only two length-scales apply: the infinitesimal scale of a material point and the macro-scale size of the entire specimen, structure or region. The spontaneous appearance of localized features suggests intermediate scales that are only accessible with more complex modeling.

4.7.1. Force chains

Force chains are strings of particles that touch at contacts that bear larger-than-average forces and trend in a sinuous but roughly linear manner through a granular material. These chains are typically oriented in the direction of the major principal compressive stress. Force chains, one of the most striking features of granular stress, are conspicuously revealed in photo-elastic experiments and in simulations. Evidence of force chains was first found in photo-elastic experiments on powdered glass particles that were immersed in a fluid that matched the refractive characteristics of glass [WAK 58]. Photographs revealed irregular fibrous, filamentary structures trending in consistent directions through the material. Confirmation of force chains was even clearer in experiments on photo-elastic rods by Dantu [DAN 57], on photo-elastic disks by Drescher and de Josselin de Jong [DRE 72] and on photo-elastic ovals by Oda and his co-workers [ODA 82, KON 82]. The transfer of internal force from one disk to another in inter-connected sequences of highly loaded disks is quite conspicuous in their photographs. Force chains were also plainly displayed in the early DEM results of Cundall [CUN 79, CUN 82], who first used the term, and of Thornton and Barnes [THO 86].

Figure 4.25 shows the normal contact forces among the particles of a disk assembly, with each line's thickness representing a contact's force magnitude. Sinuous force chains are seen to trend in the vertical direction, coincident with the direction of biaxial compression, through the height of the assembly and across the upper and lower periodic boundaries. Force chains are a ubiquitous feature during deviatoric loading: at the start of loading, at the peak and critical states, and within and outside of shear bands. Force chains of as many as 15–20 particles have been reported for two-dimensional assemblies, and the probability distribution of their length has been shown to be roughly exponential [PET 05, SUN 10]. The chains of force occur primarily among an assembly's strong contacts (section 4.5). Peters *et al.* [PET 05] have shown, however, that the network of strong contacts does not fully coincide with the set of force chains, although approximately half of the particles in the strong network are part of force chains. The lateral contacts alongside the force chain particles are primarily weak contacts, and contact slip usually occurs among this weak network of contacts rather than within the strong contacts of the force chains.

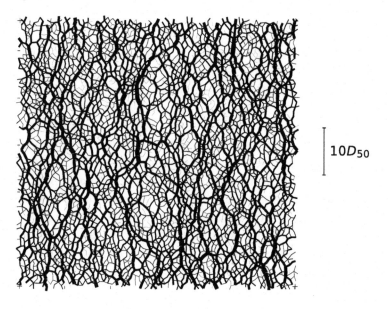

$10D_{50}$

Figure 4.25. *Spatial distribution of contact forces in an assembly of 4096 disks. The assembly was compressed vertically while maintaining constant mean stress. Each line joins the centers of two particles and has a thickness in relation to the normal contact force. The vertical strain, $\varepsilon = 2.5\%$, is at the peak stress state*

Although force chains are easily seen in displays of contact force, such as Figure 4.25, quantifying the length, intensity and longevity of force chains requires an algorithm for identifying their presence and location. Several methods have been proposed for identifying force chains, each with its own implied definition of a chain. These definitions include criteria on a chain's length, orientation and linearity [ZHA 01, PET 05]. A less biased method assumes that a force chain is simply a sequence of highly loaded contacts, without regard to length or orientation.

The adjacency matrix $\mathbf{A}^{adj}$ was introduced in section 1.2.2 for describing the topology (connectivity) of the particles and contacts within a granular material. We can identify force chains by using a *weighted* particle graph in which each edge (contact) is assigned a scalar weight: either its inverse normal force, $1/f^{pq,n}$, or its inverse force magnitude, $1/|\mathbf{f}^{pq}|$. For non-convex particles, this weight is derived from the sum of the forces at the multiple contacts between a particle pair (Figure 1.6(c)). A weighted adjacency matrix $\mathbf{F}^{adj}$ is then populated with these inverse contact forces for the contacting particles but with ∞'s placed in other matrix locations. By replacing the 0's in $\mathbf{A}^{adj}$ with ∞'s, the Floyd-Warshall algorithm can be used to rapidly find the distance matrix $\mathbf{A}^{dist}$ (discrete metric), in which the element A_{PQ}^{dist} is the number of contacts that separate particles P and Q. If the two particles are touching, then $A_{PQ}^{dist} = 1$, whereas particles that are indirectly touching through a chain of intermittent particles have a distance greater than 1 (see Figure 4.9(a)). In this manner, row P (or column P) of $\mathbf{A}^{dist}$ gives the distance between particle P and every particle in an assembly. The Floyd-Warshall algorithm can also be used to compute a "nearest weighted distance" matrix $\mathbf{F}^{dist}$, in which each element F_{PQ}^{dist} is the sum of the inverse forces, $1/f^{pq,n}$, of all contacts along the contact chain between the two particles P and Q. A chain of large forces between P and Q is indicated by a small value of F_{PQ}^{dist}. If there are N particles in an assembly, we can characterize the intensities of the contact forces along the chains of contacts between each of the $N(N + 1)/2$ pairs of particles, by considering the weighted harmonic mean $\overline{f}_{PQ}$ of the normal contact force between any pair, P and Q:

$$\overline{f}_{PQ} = A_{PQ}^{dist}/F_{PQ}^{dist} = A_{PQ}^{dist} \left(\sum_{pq \in PQ} 1/f^{pq,n} \right)^{-1} \qquad [4.27]$$

where $pq \in PQ$ are the contacting pairs in the chain between particles P and Q (no summation is implied with repetition of the latter indices). By ranking the pairs P and Q by their force intensity $\overline{f}_{PQ}$ and then by length A_{PQ}^{dist}, we can identify the dominant force chains within an assembly [KUH 16a].

Force chains are ephemeral features with a fairly short lifespan. The author compared the force chains at various strains in three-dimensional assemblies of spheres and sphere-clusters undergoing triaxial compression [KUH 16a]. No correlation was found between the force chains that were present at the strains of 6% and 7%, indicating that the force-chain network was almost entirely rerouted across the strain difference of 1%. Even at strains separated by 0.1%, the network had been largely transformed into a new network. Tordesillas *et al.* [TOR 09] have shown that abrupt changes in the force-chain network coincide with the stress drops that are observed in experiments and simulations (section 4.1.1), which are accompanied by increases in the kinetic energies of the particles and changes in the distributions of normal and tangential contact forces. These events are often characterized as "force-chain buckling". Because buckling connotes a folding or bending of structural elements, the author prefers to view them as abrupt *rerouting* or *rearrangement* events in which local stiffness-driven instabilities propel particles to new, reformed arrangements.

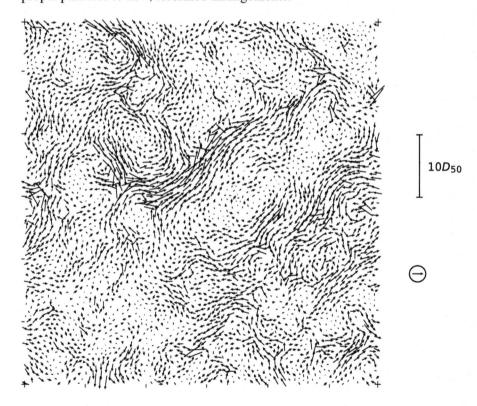

$10D_{50}$

Figure 4.26. *Circulating pattern of particle movements in a simulated disk assembly undergoing vertical biaxial compression. The particle movements are shown at a small compressive strain (vertical strain $\varepsilon = 0.2\%$) during a short strain interval $\Delta\varepsilon = 5 \times 10^{-5}$. The size of 10 median particle diameters is shown on the right, and the small (circled) arrow is a velocity three times the mean affine rate*

4.7.2. *Circulation cells*

Simulations of two-dimensional disk and ellipse assemblies consistently reveal coherent vortex, eddy-like circulation patterns in the residual, non-affine particle movements [WIL 97, MIS 98, RAD 02]. Figure 4.26 shows the residual differences $du^p - du^{p,\text{ideal}}$, between the actual and the ideal, affine particle movements in the simulation of a disk assembly undergoing biaxial compression (see equation [4.4]). The circulation patterns, termed "circulation cells" by Williams and Rege [WIL 97], have been observed in most forms of deviatoric loading (biaxial compression, simple shear and annular Couette shear), under various boundary types (rigid, flexible and periodic), and both in simulations and in laboratory studies (for example, [UTT 04, ABE 12]). These non-affine movement patterns are best discerned by magnifying the movements relative to the physical size of an assembly. Although they are more difficult to perceive in three-dimensional assemblies, similar patterns can be found by interpolating particle movements onto a rectangular grid and viewing the movements in cross-sections of the assembly [ABE 12].

Circulation cells appear at the very start of loading (zero strain), continue throughout deviatoric loading and are even present within shear bands [KUH 02, THO 06, ALO 06]. Circulating motions appear to be most intense among particles that comprise the weak force network (see section 4.5) [TOR 08]. In disk simulations, the average number of particles within a circulation cell increases with increasing strain, from a few tens of particles at the start of loading to 100 or more particles at larger strains. After shear bands have formed, circulation zones can encompass large spatial portions of the material outside of the narrow shear band, even as smaller circulation cells occur within the shear bands themselves [THO 06] (section 4.7.4). The two-dimensional sections of Abedi *et al.* [ABE 12] of sand grain motions within a shear band show circulation cells with radii as large as 3–4 times the median particle size. Although no comprehensive study has been made on their longevity or temporal patterning, circulation cells are non-persistent, ephemeral patterns: vortices present at the start of loading seem to relocate and reorganize during the first 1% of strain and continue to reorganize and increase in size until persistent shear bands have formed. Within a shear band, Abedi *et al.* [ABE 12] observed that circulation movements alternate between periods of 0.2–0.4% strain in which they are entirely absent and longer periods of intense circulation activity.

4.7.3. Micro-bands

As with particle movements, deformation within a granular assembly is highly heterogeneous and is spatially organized. The patterning of local deformation within a two-dimensional assembly can be readily detected by computing the displacement gradient $\mathbf{U}^L$ within each void polygon L or within each void triangle of a Delaunay tessellation using the methods of sections 1.4.1.1–1.4.1.2. These local deformations can then be filtered with a particular template $\mathbf{\Phi}$ to compute a scalar measure ϕ^L of the intensity of this type of deformation:

$$\phi^L = \mathbf{U}^L : \mathbf{\Phi}/|\mathbf{\Phi}| \qquad\qquad [4.28]$$

where the inner produce $\mathbf{U}^L : \mathbf{\Phi} = U_{ij}^L \Phi_{ij}$ and $|\mathbf{\Phi}| = \sqrt{\Phi_{ij}\Phi_{ij}}$. That is, ϕ^L is the magnitude of the projection of the local deformation $\mathbf{U}^L$ onto the deformation mode $\mathbf{\Phi}$. These modes can include the dilation of a void and the average rotation of the cluster of particles that surrounds a void. A particularly revealing deformation mode is plotted in Figure 4.27(a) [KUH 99]. The plot is taken from a simulation of vertical biaxial compression of a dense assembly of disks, and a strong patterning of deformation is found when the filter $\mathbf{\Phi}$ corresponds to the slip modes shown in Figure 4.27(a):

$$\mathbf{\Phi}^{\beta\pm} = \begin{bmatrix} \cos\beta\sin\beta & \mp\cos^2\beta \\ \pm\sin^2\beta & -\cos\beta\sin\beta \end{bmatrix} \qquad\qquad [4.29]$$

Micro-bands, thin zones of intense deformation, are seen to trend downward and to the right when the right-slip template $\mathbf{\Phi}^{\beta^-}$ is applied (Figure 4.27(b)). Although not shown, a corresponding pattern of micro-bands trends downward and to the left when the left-slip template $\mathbf{\Phi}^{\beta^+}$ is used. These features are also recognized by plotting the intensity of the deviatoric strain – the norm of the local displacement gradients $\mathbf{U}^L$ – after subtracting the mean strain [LAN 01].

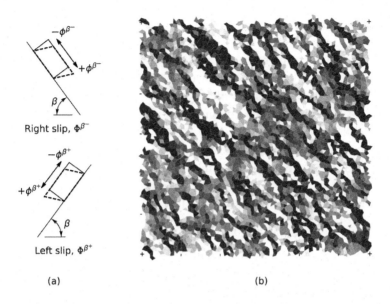

(a)   (b)

Figure 4.27. *Micro-bands: (a) right-slip and left-slip deformation modes in equation [4.29] and (b) right-slip micro-bands in a simulated disk assembly undergoing vertical biaxial compression [KUH 99]. The deformation rate is shown at the small strain $\varepsilon = 0.2\%$ during a short strain interval of $\Delta\varepsilon = 5 \times 10^{-5}$. The darkness of each void polygon corresponds to the direction and intensity of the right-slip deformation ϕ^L, with $\beta = 50°$. The darkest regions are deforming in a right-slip mode at a rate twice the mean affine rate. The snapshot corresponds to that in Figure 4.26*

Micro-bands are revealed in simulations with disk assemblies [KUH 99, ROU 03, ALO 06] and in physical experiments with wood, aluminum and plastic rods [CAL 97, MIS 98, LAN 01, HAL 10]. These micro-bands are thin chains of void cells that trend oblique to the principal directions of the bulk strain and within which slip deformation is most intense. Because of their ubiquity during deviatoric deformation, they can be considered the kinematic counterpart of force chains, which also reliably appear during deviatoric loading (section 4.7.1). Micro-bands are separated by thicker zones within which the slip deformation Φ^β is less than the assembly mean. These features appear at the very start of loading, at which the intensities $\phi^{\beta\pm}$ are subtle and the thickness ranges between $1\frac{1}{2}$ and $2\frac{1}{2}$ particle diameters. At larger strains, micro-bands become more intense (expressing greater deformation heterogeneity), thicker (with thicknesses of up to four particle diameters) and trend in an oblique direction that is somewhat closer to the direction of the compressive strain. With disk assemblies, micro-bands can even be seen to emerge and dissipate within shear bands, which are much thicker features than the micro-bands themselves, and such micro-bands are typically located between co-rotating circulation cells [TOR 08] (note, however, that micro-bands have not been observed within the shear bands of sand specimens [ABE 12]). Micro-bands are transient features, usually emerging and dissipating in spans of less than 0.2% of bulk strain, and they usually emerge among the larger voids rather than in the smaller, triangular voids (3 cycles) of two-dimensional materials. Dilation tends to be more intense within micro-bands [KUH 99], so that these features can be recognized as thin dilation zones when the local dilation has been resolved at the particle scale [HAL 10].

Although micro-bands are most easily discerned in two-dimensional simulations and experiments, they also occur during the deviatoric loading of three-dimensional assemblies. To distinguish such micro-bands, we must interpolate the particle displacements onto a regular three-dimensional grid (e.g. as in section 1.4.2.3). In this manner, micro-band features can be identified by either computing the deformations within the voxels or using Fourier analysis techniques.

4.7.4. *Shear bands*

Of the localization phenomena discussed in this section, shear bands were the first to be reported and have received the greatest attention. Whereas force chains, micro-bands and circulation cells are only revealed with special experimental techniques or with computer simulations, shear bands are clearly expressed in standard geotechnical tests and have been an important topic of geomechanics for over fifty years. An understanding of the origin and micro-structure of shear bands has largely been gained from experimental works during the past two decades. DEM simulations have revealed additional information, but difficulties in creating and analyzing large assemblies have, even now, limited their utility. And even after intense attention, the cause, the nature and the very definition of shear bands continue to raise important questions.

To start, the definition of a shear band can be either narrow or broad. Shear bands are most clearly expressed as roughly linear (flat) and thin zones of intense localized shearing deformation that are relatively stationary and separate neighboring regions of minimal deformation or regions that deform in directions different from that within the band. As such, they are readily identified in conventional triaxial and plane-strain tests that employ rubber membranes: shear bands are expressed as a wrinkling or deformation of the membrane. Unless special measurement techniques are applied, such visual evidence requires an endurance of the band for several percent of strain, which is the defining characteristic of a *persistent shear band*. Other localization features share the primary features of persistent shear bands (linear, thin and stationary) but are ephemeral features, and these non-persistent or *temporary shear bands* usually appear before and while a persistent band is being formed. Although temporary shear bands are readily identified in simulations, their fleeting presence is more difficult to capture in a laboratory setting.

Because both persistent and temporary shear bands have physical dimensions that are much larger than the particle size but are considerably smaller than the specimen size, their existence requires a more refined interpretation of laboratory data. Laboratory apparatus provide forces and movements at the boundaries, which express the average stress and deformation within the specimen's volume. As will be seen, the large size of a shear band, when compared with the particle size, and the large disparity of deformations inside and outside the band can call into question the very meaning of material behavior, which is conventionally viewed as a local characteristic that is possessed by material points within an ideal continuum.

The interpretation of laboratory and simulation data is necessarily influenced by the continuum model for which the data is being considered. Four interpretations of a shear band have been expounded (Figure 4.28). Early models viewed shear bands as two-dimensional surfaces (failure planes or surfaces) of zero thickness, across which displacements and strains are discontinuous (Figure 4.28(a)). This interpretation of shear bands is consistent with the limit-equilibrium, slip-surface modeling of landslides, retained backfills and foundation soils. In the bifurcation framework of Thomas [THO 61], Hill [HIL 62] and Mandel [MAN 62], shear bands are viewed as regions of finite thickness within which the strain rate is uniform but differs from that of the surrounding region (Figure 4.28(b)). The conditions necessary for a jump in strain between the two regions are posed as a bifurcation problem in which the solution corresponds to a singular acoustic tensor of the continuum stiffness [RIC 76]. More recent evidence shows that the strain within a shear band, when averaged along the length and width of the band, varies smoothly across the band's thickness and smoothly merges into the strain outside of the band (Figure 4.28(c)). This evidence is the basis of continuum models that introduce a length-scale and are intended to capture smooth meso-scale transitions of strain: for example, by employing generalized continua (e.g. Cosserat continua) or by using a continuum stiffness that depends on the first (and perhaps higher) spatial gradients of strain. The most recent evidence, gained largely from discrete element (DEM) simulations and high-definition laboratory observations, reveals elaborate grain-scale patterns of movement and grain-to-grain mechanisms within a band that operates to resist deformation while confining the most active movements to the narrow shear zone (Figure 4.28(d)). This view has led to micro-scale models based on hypothetical mechanisms, such as "force-chain buckling" and "force transmission bottlenecks" [TOR 07, TOR 15]. Rather than expounding on these various models – macro-scale, meso-scale and micro-scale – this section is limited to descriptive information gained from laboratory tests and computer simulations.

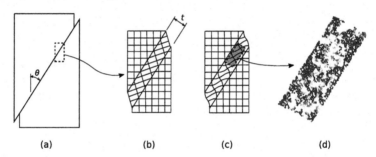

(a)   (b)   (c)   (d)

Figure 4.28. *Four interpretations of a shear band: (a) as a slip surface of discontinuous displacement; (b) as an abrupt transition zone of different strain; (c) as a continuous transition of strain within and across the band boundaries; (d) as a zone of intense but coordinated particle movements (shading shows the rate of particle rotations within a shear band). The choice of interpretation depends on the continuum framework in which the band is viewed and the scale at which movements are resolved*

4.7.4.1. *Detecting shear bands*

The presence and evolution of shear bands have been revealed with a number of experimental and simulation techniques. To begin, shear bands can be inferred from abrupt changes in the boundary stresses and movements during standard strength tests, particularly with materials of medium and high density, which tend to soften and dilate at strains beyond the peak state. Figure 4.29 represents typical results of a large dense sand specimen during drained plane-strain biaxial compression, as reported by a number of investigators [HAN 93, FIN 97b] (testing with other boundary conditions has also been reported, with similar results [DES 85, SAA 99, WAN 01]). The abrupt reduction of deviator stress q at the peak state suggests that deformation, which was originally occurring in a more homogeneous fashion throughout the specimen, has been localized within the small volume of a softening region – the shear band – producing only small movements at the boundaries. The abrupt change in the rate of volume change, as measured by the boundary movements or by the changing volume of the pore fluid (lower part of Figure 4.29(a)), indicates that dilation, previously occurring throughout the specimen, has become restricted to the small volume of a persistent shear band, which soon reaches the critical state. The apparatus for plane-strain biaxial compression testing can permit measurement of the horizontal movements of the top or bottom platens (sleds) or movements along the expanding sides of the specimen (usually along rubber membranes), and any abrupt changes in these movement rates are also direct indications of a concentration of deformation within shear zones, a change in the direction of the material's deformation or a discontinuity in the spatial pattern of deformation [DES 04]. Although the presence of shear bands can be inferred from such macro-scale results, other techniques are required to determine the number, location, orientation and thickness of shear bands. Furthermore, movements and forces at the boundaries can only indicate those shear bands that have persisted across a sufficiently large range of strains, but they might not resolve the presence of non-persistent bands (whether pre-peak or post-peak) and cannot reveal the evolution of the strain field within a persistent band. Figure 4.29(b) represents results of a DEM simulation of shear band formation in a relatively small assembly of 120,000 spheres [SUN 13]. Because the shear band volume comprises a significant fraction of the specimen's volume (about 12%), the bulk behavior differs from that of a physical specimen that is much larger than its relatively thin shear zones (e.g. Figure 4.29(a)). During the broader stress peak of the simulation in Figure 4.29(b), multiple non-persistent, temporary shear bands were active, and bulk softening occurred only after a single band had become persistent. The difference in behaviors of Figures 4.29(a) and (b) show that the average bulk behavior depends on specimen size and is further evidence that an intermediate length-scale applies to granular behavior.

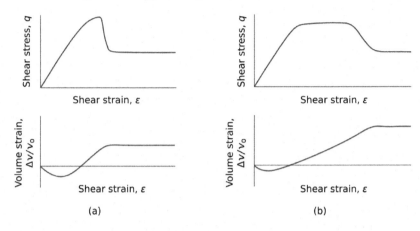

(a)   (b)

Figure 4.29. *(a) Typical results for a plane-strain biaxial compression test, showing the abrupt change in stress and volumetric strain on the formation of a persistent shear band (e.g. [ALS 03, DES 04]) and (b) results of a simulation of plane-strain simple shear of a small DEM specimen [SUN 13]*

Shear bands are also detected by observing surface movements across the boundary membrane or among the boundary particles. A number of investigators have painted grid lines onto the surface of the membrane prior to plane-strain biaxial [DES 85, TAT 90, HAN 91, ALS 00], conventional triaxial [ALS 03], true-triaxial [DES 85] and hollow-cylinder loading tests [SAA 99]. The pattern of grid deformation indicates the location and inclination of the shear bands and gives a rough estimate of band thickness [ODA 98b]. Movements within a specimen's interior have been measured by embedding a grid of lead tracer particles within the specimen and tracking their displacements with X-ray imaging [ART 82, SCA 82]. Sequences of surface photographs have also be used to track particle movements and observe the evolution of any localized deformation during loading. In this manner, the spatial resolution of a shear band's location, shape and thickness have been gained with stereo-photographic methods in which two photographs, taken at different strains, are optically compared, so that the surface expression of a shear band appears as a topographic relief [FIN 97a, DES 04]. High resolutions are achieved by spraying a speckle pattern onto the membrane or by photographing the surface particles themselves. Digital image correlation (DIC) is a further refinement, in which particle movements are tracked by identifying individual grains in high-resolution digital images of a deforming soil specimen [REC 06, REC 10].

Imaging techniques that detect the expression of a shear band on the surface of a sample have been augmented by imaging and sectioning techniques that measure density changes within a specimen. Such methods are particularly revealing with conventional triaxial compression tests, since interior shear banding or other localization patterns might not be expressed on a specimen's surface. Gamma-ray and X-ray absorption methods, combined with computed tomography, reveal density change within triaxial specimens and have exposed elaborate patterns of multiple radially symmetric shear bands that form during post-peak strains [DES 96, ALS 03].

Shear bands have also been studied with computer simulations. Simulations have provided important information about the distribution of strain and stress within shear bands, the formation and evolution of shear bands on small temporal and spatial scales, grain-scale movements and contact forces and their patterning within a shear band and the concurrent deformations that occur outside of a shear band. These studies include both two-dimensional [CUN 89b, BAR 91, KUH 05a, ZHA 13a] and three-dimensional simulations [FAZ 06, SUN 13].

4.7.4.2. Formation and evolution

As noted in section 4.7.3, thin ephemeral zones of relatively intense shearing occur spontaneously at the start of loading. The shear strains within these micro-bands become more intense with increasing strain, and they become steeper (relative to directions of bulk extension) and thicker. Whether micro-bands are the precursors of shear bands has not yet been determined, but non-persistent shear bands that are thicker than micro-bands are known to form well before the peak stress, and non-persistent bands have been observed in experiments (primarily in plane-strain biaxial compression tests [DES 90a, FIN 97a, DES 04]) and in simulations [KUH 05a, THO 06, SUN 13]. Multiple non-persistent shear bands usually form in conjugate directions (i.e. with orientations that are symmetric with respect to the principal direction of compression) during the strain-hardening that occurs prior to the peak stress [REC 06]. A non-persistent band can alternate between an active and dormant condition, as localization shifts back and forth among several non-persistent bands, which are in seeming competition to attract an assembly's deformation [TAT 90, KUH 05a]. In undrained biaxial tests, the formation of multiple shear bands in conjugate directions can continue to occur at large strains, such that a single persistent shear band is never fully expressed [HAN 91, HAR 95]. If the loading and boundary conditions permit and if a dominant shear band has formed at or near the peak stress, subsequent shearing deformation occurs almost exclusively within the persistent band, and the softening (stress reduction) that occurs with advancing strain within the band is accompanied by unloading (strain reversal) in regions outside of the band.

The inception and location of the eventual persistent band is influenced by the boundary conditions (walls, membranes, etc.), the aspect ratio of the specimen and the presence of non-homogeneous conditions (so-called imperfections) within the specimen or along its boundaries. Material softening within a persistent band produces unloading at the locations of previous non-persistent bands, subduing their further activation. As such, the question of whether a persistent shear band is initiated before, at, or after the peak stress is somewhat misleading, as the eventual persistent band has likely been expressed at earlier (and perhaps at multiple) strains as a temporary, fleeting feature. As strain begins to concentrate within a single band, this zone of intense localization grows in length and width, perhaps extending across the specimen's dimensions. A band can be "reflected" at the boundary in its complementary direction, so that the two bands intensify as complementary pairs [DES 90a]. Alternatively, a complementary band can spontaneously appear much later in the deformation process [ALS 00].

Most experimental studies of shear bands have been conducted as plane-strain biaxial compression tests, since shear bands readily form and are most easily observed under these conditions. In conventional axisymmetric triaxial compression tests, non-homogeneous deformation usually begins as a bulging of the cylindrical sample, and the formation of shear bands is suppressed until larger strains. Significant localization may not appear in triaxial specimens until strains well beyond the peak stress, at strains of 20% or more [DES 96]. Whether and when a shear band occurs under triaxial conditions depends, in part, on details of the boundary conditions and specimen preparation, and conditions that promote material inhomogeneity or imperfect loading (e.g. large specimen height-to-width ratios, frictional platens, tilting platens, etc.) tend to hasten the onset of localization.

4.7.4.3. *Thickness and orientation*

Shear bands are not two-dimensional slip planes, but are three-dimensional zones of localized deformation. Their thickness is closely related to particle size, and thicknesses of 10–30 times the median particle size D_{50} have been reported [ROS 70, FIN 97b, VAR 98]. This large range in the thickness–size ratio, t/D_{50} (Figure 4.28(b)), is due to the dependence of the ratio on particle size, specimen preparation method, test conditions, particle shape and measurement method. The ratio has been shown to vary as follows: the ratio decreases significantly with increasing D_{50} [YOS 97, ALS 99, DES 04, REC 11]; [FIN 97a, ALS 99, DES 04]; decreases with increasing mean stress [DES 04]; decreases with increasing particle non-sphericity or contact rolling resistance [ZHA 13a]. Because the shearing strain varies across the thickness of a shear band and the band can be wavy along its length and width [ODA 98b], any measurement of thickness necessarily involves an interpretation of a band's periphery and an averaging of the thickness across the length and width of the band.

Shear bands are oblique to the principal stress directions. Although they are commonly thought to be flat, planar zones, bands can be slightly curved, particularly when influenced by the presence of hard platen surfaces [ODA 98b, DES 04]. Two reference directions are commonly used for comparing the angle θ between the major principal stress and the general plane of a shear band (Figure 4.28(a)). The Mohr-Coulomb failure criterion is based on the peak angle ϕ of bulk internal friction – the angle in equations [4.1] and [4.2] of maximum obliquity of the shear stress relative to the normal stress – and an assumption that this obliquity is realized along the shear band is the basis of the Coulomb orientation angle, $\theta_C = \frac{\pi}{4} - \frac{\phi}{2}$. Roscoe's kinematic assumption of zero extension along the length and width of a shear band leads to an orientation $\theta_R = \frac{\pi}{4} - \frac{\psi}{2}$, where ψ is the dilation angle (see equation [4.3]) that is in effect when the shear band forms [ROS 70]. Although the Coulomb and Roscoe orientations are theoretical results derived from different interpretations of the failure condition, both assume that stress and strain are coaxial within a shear band (i.e. with aligned principal directions) and that strength and dilation within a shear band are the same as those in a material undergoing homogeneous deformation. Based on experimental evidence that shear band orientation usually lies between the Coulomb and Roscoe angles, Arthur *et al.* [ART 77b] proposed an alternative reference orientation, $\theta_A = \frac{\pi}{4} - \frac{1}{2}(\frac{\phi}{2} + \frac{\psi}{2})$. Most plane-strain biaxial compression tests yield angles θ that are near θ_A and are closer to θ_C than to θ_R [BAR 90, DES 04]. Moreover, the measurements of Saada *et al.* [SAA 99] contradict the assumption, inherent in the angle θ_R, of zero extension along the shear band. Persistent shear bands are usually steeper (with smaller θ) than the temporary shear bands from which they had formed [TAT 90, HAR 95, FIN 97a]. The orientation angle θ of persistent shear bands increases with an increase in the mean stress [HAN 93, YOS 97, DES 04], increases with an increase in particle size [ALS 00, DES 04] and is larger for loose specimens than for dense specimens [DES 04].

4.7.4.4. *Inside shear bands*

Although bifurcation theories, when applied to simple classical continua, suggest that stress and deformation change abruptly as one passes into a shear band from regions outside of the band, simulations have shown that this is not the case. Figure 4.30 shows the results of a simulation of simple shear of a large assembly of more than 120,000 spheres, in which the top surface was displaced horizontally to the right (in the x_1 direction) relative to the bottom surface [SUN 13]. The persistent shear band, which fully developed at a shear strain of 6%, extends horizontally through the width and thickness of the specimen and passes across the periodic boundaries along the specimen's vertical sides. Vertical profiles along the x_2 height are shown for the particle movements u_1^p and the accumulated shear strain $\varepsilon_{12}/\langle\varepsilon_{12}\rangle$ that had occurred between the start of loading and a shear strain of 12% (Figures 4.30(a) and (b). The figure also shows the final density, as void ratio e, at the 12% strain (Figure 4.30(c)). A shear band is clearly present between the heights of about 83–97 mm, a thickness that is about 14 times larger than the median sphere diameter of 1 mm. The plots of strain and void ratio were developed by averaging these quantities across the length of the band and across its transverse width, noting that the band encompasses several thousands of particles (the method described in section 1.4.2.1 was used to compute the shear strain of Figure 4.30(b)). The smooth transition of density and strain across the x_2 thickness of the shear band points to the difficulty in assigning a precise thickness to a band. Because the material was dense prior to shearing, the void volume increased during loading and is greater within the band than outside. The void ratio (Figure 4.30(a)) is seen to transition, however, from a value of about 0.54, well outside of the band, to a maximum of about 0.70 near the middle of the shear band. A similar transition of void ratio within a shear band is also seen in the results of Alshibli and Sture [ALS 99].

Although the band thickness is identified as 14 mm on the basis of the shear strain profile (Figure 4.30(b)), regions both above and below this thickness have attained void ratios that are above the 0.54 value, and we might interpret a thickness of about 20 mm on the basis of density alone.

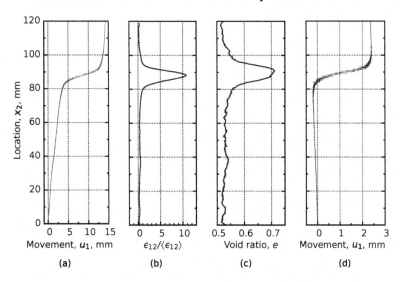

Figure 4.30. *Results of a DEM simulation of horizontal simple shearing of a large assembly of spheres. The vertical profiles show transitions within a persistent shear band: (a) horizontal (shearing) particle movements between the strains of 0% and 12%; (b) accumulated shear strain between the strains of 0% and 12%; (c) final void ratio at 12% strain; (d) horizontal particle movements between the strains of 6% and 8% (see [SUN 13])*

Figure 4.30(b) reveals the subtle remnants of non-persistent shear bands that had appeared before the shear strain of 6%, seen as the slight changes in strain and density at heights x_2 of 5–15 mm and 32–42 mm, although these features are faint in comparison to the dominant persistent shear band, which is prominent in the accumulated strain of Figure 4.30(b). Between the strains 2% and 6%, these non-persistent bands were of similar intensity to the still-forming band that eventually became persistent at 6% strain.

Figure 4.30(d) shows the particle movements that had occurred between strains of 6% and 8%, after the persistent shear band had fully formed and during the period of strain softening. We can recognize the reversal of strain above and below the band as a slightly reversed slope, which occurs even as large strains are accumulating within the shear band.

The void ratio within a shear band can be quite large and can exceed the maximum value obtained with standard laboratory methods for preparing reconstituted sand samples (the value e_{max} described in section 4.1). By using optical measurements of petrified thin sections, Oda and Kazama [ODA 98b] found that void ratios within a shear band were greater than 1.00 for two sands, whose e_{max} values were 0.973 and 0.96. The thin sections also revealed that the voids were highly elongated in the direction perpendicular to (and through the thickness of) the shear band. Two-dimensional experiments with oval photo-elastic particles [ODA 82, ODA 98b] and simulations with disks in which a rotational resistance is embedded in the contact model [IWA 98, TOR 08] have shown that elongated voids are located between columns of particles that trend through the thickness of a shear band. These columns of particles lean and rotate as the material on both sides of (say, above and below) the shear band shifts past each other, and the columns periodically transfer their loads to nearby particles through a process of grain rearrangement.

The particle and contact fabrics are usually different for zones within and outside a shear band. After biaxial compression, Oda and Kazama [ODA 98a] found that the elongated dimensions of sand particles outside of a shear band are predominantly aligned orthogonal to the direction of compressive loading, consistent with the trends in sections 1.2.4 and 4.6.1. Within a shear band, however, the longer dimensions of particles tend to be aligned with the direction (length) of the shear band and oblique to the principal direction of compressive stress. Measurements of the orientations of contact normals (e.g. Figures 4.20(a) and 4.23) after the simulated biaxial shearing of sphere assemblies show that the normals are preferentially oriented in the direction of the major principal compressive stress, both inside and outside a shear band. The predominant orientations do differ, however, by about 4–5° for the two regions [SUN 13].

Simulations and laboratory tests have also revealed even finer details of the movements and forces within shear bands. Using digital image correlation techniques to resolve the movements of sand particles along transparent side walls during biaxial compression, Rechenmacher and Abedi discovered vortex structures within shear bands [REC 11]. Within the middle portion of a shear band, circulation cells similar to those in section 4.7.2 rotate in the same direction as the mean vorticity, whereas counter-rotating circulation cells develop near the peripheries of the band's thickness. Similar patterns have been observed in simulations of disk particles [ALO 06, TOR 08]. By resolving deformation at a small scale, Rechenmacher *et al.* [REC 10] also noted that both dilation and contraction occur within a shear band in alternating zones along the length of a band. By measuring the displacements of a painted grid, Tatsuoka *et al.* [TAT 90] have shown a similar alternating pattern of more and less intense shear strains along the length of a band.

Permissions

1) Figure 1.1 is from Krumbein W. C., Sloss L. L., *Stratigraphy and Sedimentation*, W.H. Freeman and Company, San Francisco, 1963.

2) Figure 1.8 is republished with permission of Elsevier BV, from M. R. Kuhn, "Structured deformation in granular materials," *Mechanics of Materials*, 31(6):407–429, 1999; permission conveyed through Copyright Clearance Center, Inc.

3) Figure 2.4 is republished from T. W. Lambe and R. V. Whitman, *Soil Mechanics*, John Wiley, New York, 1969. Copyright © 2014 John Wiley.

4) Figure 2.6 is republished with permission of The Royal Society of London, from K. L. Johnson, "Surface interaction between elastically loaded bodies under tangential forces," *Proceedings of the Royal Society of London, A*, 230(1183):531–548, 1955; permission conveyed through Copyright Clearance Center, Inc.

5) Figure 2.8 is republished from M. R. Kuhn, "Transient rolling friction model for discrete element simulations of sphere assemblies," *Comptes Rendus Mécanique*; 342(3):129–140, 2014. Copyright © 2014 Elsevier.

6) Figure 2.10 is republished from M. R. Kuhn, "Transient rolling friction model for discrete element simulations of sphere assemblies," *Comptes Rendus Mécanique*; 342(3):129–140, 2014. Copyright © 2014 Elsevier.

7) Figure 4.14 is republished with permission of Elsevier BV, from M. R. Kuhn, "Micro-mechanics of fabric and failure in granular materials," *Mechanics of Materials*, 42(9):827–840, 2010; permission conveyed through Copyright Clearance Center, Inc.

8) Figure 4.27 is republished with permission of Elsevier BV, from M. R. Kuhn, "Structured deformation in granular materials," *Mechanics of Materials*, 31(6):407–429, 1999; permission conveyed through Copyright Clearance Center, Inc.

Notation

Item	Object notation	Index notation	Example, description
Scalar	c or α	c or α	Void ratio, volume
Vector	$\mathbf{c}$ or $\boldsymbol{\alpha}$	c_i or α_j	Velocity
Tensor	$\mathbf{C}$ or $\boldsymbol{\Gamma}$	C_{ij} or Γ_{jk}	Stress
Vector, m rows	$[\mathbf{c}]$ or $[\mathbf{c}]_{m\times 1}$	c_i	Vector $[\boldsymbol{\Delta}]$, Eq. (1.124)
Matrix, m rows, n columns	$[\mathbf{C}]_{m\times n}$ or $[\mathbf{C}]$	C_{ij}	Matrix $[\mathbf{H}]$, Eq. (3.44)
Identity, Kronecker tensor	$[\mathbf{I}]$ or $\boldsymbol{\delta}$	I_{ij} or δ_{ij}	$I_{ij},\ \delta_{ij} = \begin{cases} 0 & i \neq j \\ 1 & i = j \end{cases}$
Vector inner product	$\mathbf{a} \cdot \mathbf{b}$	$a_i b_i$	$\sum_i a_i b_i$
Outer, cross product	$\mathbf{a} \times \mathbf{b}$	$e_{ijk} a_j b_k$	$\mathbf{e}_1 = \mathbf{e}_2 \times \mathbf{e}_3$
Dyad	$\mathbf{a} \otimes \mathbf{b}$	$a_i b_j$	Eq. (1.46)
Matrix inner product	$\mathbf{A} : \mathbf{B}$	$A_{ij} B_{ij}$	$\sum_i \sum_j A_{ij} B_{ij}$
Operator, matrix product	$\mathbf{C} \cdot \mathbf{a}$ or $[\mathbf{C}][\mathbf{a}]$	$C_{ij} a_j$	$\sum_j C_{ij} a_j$
Operator, matrix product	$\mathbf{A} \cdot \mathbf{B}$ or $[\mathbf{A}][\mathbf{B}]$	$A_{ik} B_{kj}$	$\sum_k A_{ik} B_{kj}$
Trace	$\text{tr}(\mathbf{C})$	C_{ii}	$C_{11} + C_{22} + \ldots$
Divergence	$\vec{\nabla} \cdot \mathbf{C}$	$C_{ji,j}$	$\partial C_{ji}/\partial x_j$, Eq. (1.44)
Inverse	$\mathbf{A}^{-1}$ or $[\mathbf{A}]^{-1}$	A_{ij}^{-1}	$[\mathbf{A}]^{-1}[\mathbf{A}] = [\mathbf{I}]$
Adjoint or transpose	$\mathbf{A}^{\mathrm{T}}$ or $[\mathbf{A}]^{\mathrm{T}}$	A_{ij}^{T}	$A_{ij}^{\mathrm{T}} = A_{ji}$

Bibliography

[ABE 12] ABEDI S., RECHENMACHER A.L., ORLANDO A.D., "Vortex formation and dissolution in sheared sands", *Granular Matter*, vol. 14, no. 6, pp. 695–705, 2012.

[ADL 92] ADLER P.M., *Porous Media: Geometry and Transports*, Butterworth-Heinemann, Boston, 1992.

[AGN 07a] AGNOLIN I., ROUX J.-N., "Internal states of model isotropic granular packings. I. Assembling process, geometry, and contact networks", *Physical Review E*, vol. 76, p. 061302, 2007.

[AGN 07b] AGNOLIN I., ROUX J.-N., "Internal states of model isotropic granular packings. III. Elastic properties", *Physical Review E*, vol. 76, p. 061304, 2007.

[AI 11] AI J., CHEN J.-F., ROTTER J.M. *et al.*, "Assessment of rolling resistance models in discrete element simulations", *Powder Technology*, vol. 206, no. 3, pp. 269–282, 2011.

[ALO 06] ALONSO-MARROQUIN F., VARDOULAKIS I., HERRMANN H.J. *et al.*, "Effect of rolling on dissipation in fault gouges", *Physical Review E*, vol. 74, no. 3, p. 031306, 2006.

[ALO 08] ALONSO-MARROQUIN F., "Spheropolygons: a new method to simulate conservative and dissipative interactions between 2D complex-shaped rigid bodies", *Europhysics Letters*, vol. 83, no. 1, p. 14001, 2008.

[ALS 99] ALSHIBLI K.A., STURE S., "Sand shear band thickness measurements by digital imaging techniques", *Journal of Computing in Civil Engineering*, vol. 13, no. 2, pp. 103–109, 1999.

[ALS 00] ALSHIBLI K.A., STURE S., "Shear band formation in plane strain experiments of sand", *Journal of Geotechnical and Geoenvironmental Engineering*, vol. 126, no. 6, pp. 495–503, 2000.

[ALS 03] ALSHIBLI K.A., BATISTE S.N., STURE S., "Strain localization in sand: plane strain versus triaxial compression", *Journal of Geotechnical and Geoenvironmental Engineering*, vol. 129, no. 6, pp. 483–494, 2003.

[ALS 06] ALSHIBLI K., ALRAMAHI B., "Microscopic evaluation of strain distribution in granular materials during shear", *Journal of Geotechnical and Geoenvironmental Engineering*, vol. 132, no. 1, pp. 80–91, 2006.

[ALT 13] ALTUHAFI F., O'SULLIVAN C., CAVARRETTA I., "Analysis of an image-basedased method to quantify the size and shape of sand particles", *Journal of Geotechnical and Geoenvironmental Engineering*, vol. 139, no. 8, pp. 1290–1307, 2013.

[AND 12] ANDRADE J.E., LIM K.-W., AVILA D.F. *et al.*, "Granular element method for computational particle mechanics", *Computer Methods in Applied Mechanics and Engineering*, vol. 241–244, pp. 262–274, 2012.

[ANN 93] ANNIC C., BIDEAU D., LEMAÎTRE J. *et al.*, "Geometrical properties of 2D packings of particles", in THORNTON C. (ed.), *Powders & Grains 93*, A.A. Balkema, Rotterdam, 1993.

[ANT 04] ANTONY S.J., KUHN M.R., "Influence of particle shape on granular contact signatures and shear strength: new insights from simulations", *International Journal of Solids and Structures*, vol. 41, no. 21, pp. 5863–5870, 2004.

[ART 72] ARTHUR J.R.F., MENZIES B.K., "Inherent anisotropy in a sand", *Géotechnique*, vol. 22, no. 1, pp. 115–128, 1972.

[ART 77a] ARTHUR J.R.F., CHUA K.S., DUNSTAN T., "Induced anisotropy in a sand", *Géotechnique*, vol. 27, no. 1, pp. 13–30, 1977.

[ART 77b] ARTHUR J.R.F., DUNSTAN T., AL-ANI Q.A.J.L. *et al.*, "Plastic deformation and failure in granular media", *Géotechnique*, vol. 27, no. 1, pp. 53–74, 1977.

[ART 82] ARTHUR J.R.F., DUNSTAN T., "Rupture layers in granular media", in VERMEER P., LUGER H. (eds.), *Deformation and Failure of Granular Materials*, A.A. Balkema, Rotterdam, 1982.

[ARU 92] ARULMOLI K., MURALEETHARAN K.K., HOSSAIN M.M. *et al.*, "VELACS: verification of liquefaction analyses by centrifuge studies laboratory testing program: soil data report", Report no. Project No. 90-0562, The Earth Technology Corporation, Irvine, available at: http://yees.usc.edu/velacs, 1992.

[AST 06] ASTM, "D-2487: standard practice for classification of soils for engineering purposes (Unified Soil Classification System)", *Annual Book of ASTM Standards*, vol. 04.08, American Society for Tesing and Materials, West Conshohocken, pp. 249–260, 2006.

[AZÉ 07] AZÉMA E., RADJAÏ F., PEYROUX R. *et al.*, "Force transmission in a packing of pentagonal particles", *Physical Review E*, vol. 76, p. 011301, July 2007.

[AZÉ 09] AZÉMA E., RADJAI F., SAUSSINE G., "Quasistatic rheology, force transmission and fabric properties of a packing of irregular polyhedral particles", *Mechanics of Materials*, vol. 41, no. 6, pp. 729–741, 2009.

[AZÉ 10] AZÉMA E., RADJAI F., "Stress-strain behavior and geometrical properties of packings of elongated particles", *Physical Review E*, vol. 81, pp. 051304:1–17, 2010.

[AZÉ 13a] AZÉMA E., RADJAI F., DUBOIS F., "Packings of irregular polyhedral particles: strength, structure, and effects of angularity", *Physical Review E*, vol. 87, no. 6, p. 062203, 2013.

[AZÉ 13b] Azéma E., Radjaï F., Saint-Cyr B. *et al.*, "Rheology of three-dimensional packings of aggregates: microstructure and effects of nonconvexity", *Physical Review E*, vol. 87, p. 052205, 2013.

[BAG 96] Bagi K., "Stress and strain in granular assemblies", *Mechanics of Materials*, vol. 22, no. 3, pp. 165–177, 1996.

[BAG 04] Bagi K., Kuhn M.R., "A definition of particle rolling in a granular assembly in terms of particle translations and rotations", *Journal of Applied Mechanics*, vol. 71, no. 4, pp. 493–501, 2004.

[BAG 05] Bagi K., "An algorithm to generate random dense arrangements for discrete element simulations of granular assemblies", *Granular Matter*, vol. 7, no. 1, pp. 31–43, 2005.

[BAG 07] Bagi K., "On the concept of jammed configurations from a structural mechanics perspective", *Granular Matter*, vol. 9, nos. 1–2, pp. 109–134, 2007.

[BAR 90] Bardet J.P., "A comprehensive review of strain localization in elastoplastic soils", *Computers and Geotechnics*, vol. 10, no. 3, pp. 163–188, 1990.

[BAR 91] Bardet J.P., Proubet J., "A numerical investigation of the structure of persistent shear bands in granular media", *Géotechnique*, vol. 41, no. 4, pp. 599–613, 1991.

[BAR 92a] Bardet J.P., Huang Q., Proubet J., "A micromechanical investigation of the influence of couple stresses on failure in granular materials", in Tillerson J.R., Wawersik W.R. (eds.), *Rock Mechanics: Proceedings of the 33rd US Symposium*, A. A. Balkema, Rotterdam, 1992.

[BAR 92b] Bardet J.P., Proubet J., "Shear-band analysis in idealized granular material", *Journal of Engineering Mechanics*, vol. 118, no. 2, pp. 397–415, 1992.

[BAR 94] Bardet J.P., "Observations on the effects of particle rotations on the failure of idealized granular materials", *Mechanics of Materials*, vol. 18, no. 2, pp. 159–182, 1994.

[BAR 98] Bardet J.P., "Introduction to computational granular mechanics", in Cambou B. (ed.), *Behavior of Granular Materials*, Springer, 1998.

[BAR 01] Baraff d., "Physically based modeling: rigid body simulation", *SIGGRAPH Course Notes, ACM SIGGRAPH*, vol. 2, no. 1, 2001.

[BAT 76] Bathe K.-J., Wilson E., *Numerical Methods in Finite Element Analysis*, Prentice-Hall, Englewood Cliffs, 1976.

[BAT 90] Bathurst R.J., Rothenburg L., "Observations on stress-force-fabric relationships in idealized granular materials", *Mechanics of Materials*, vol. 9, pp. 65–80, 1990.

[BAŽ 91] Bažant Z.P., Cedolin L., *Stability of Structures: Elastic, Inelastic, Fracture, and Damage Theories*, Oxford University Press, New York, 1991.

[BEE 85] Been K., Jefferies M.G., "A state parameter for sands", *Géotechnique*, vol. 35, no. 2, pp. 99–112, 1985.

[BHA 90] Bhatia S.K., Soliman A.F., "Frequency distribution of void ratio of granular materials determined by an image analyzer", *Soils and Foundation*, vol. 30, no. 1, pp. 1–16, 1990.

[BIG 00] Bigoni D., "Bifurcation and instability of non-associative elastoplastic solids", in Petryk H. (ed.), *Material Instabilities in Elastic and Plastic Solids*, Springer-Verlag, Wien, 2000.

[BOT 13] Boton M., Azéma E., Estrada N. *et al.*, "Quasistatic rheology and microstructural description of sheared granular materials composed of platy particles", *Physical Review E*, vol. 87, no. 3, p. 032206, 2013.

[BOW 73] Bowden F.P., Tabor D., *Friction: An introduction to tribology*, Anchor Press Doubleday, Garden City, 1973.

[BRI 98] Brilliantov N.V., Pöschel T., "Rolling friction of a viscous sphere on a hard plane", *Europhysics Letters*, vol. 42, no. 5, p. 511, 1998.

[CAL 97] Calvetti F., Combe G., Lanier J., "Experimental micromechanical analysis of a 2D granular material: relation between structure evolution and loading path", *Mechanics of Cohesive-frictional Materials*, vol. 2, no. 2, pp. 121–163, 1997.

[CAT 38] Cattaneo C., "Sul contatto di due corpi elasticiti", *Accademia dei Lincei, Rendiconti, Series 6*, vol. 27, pp. 342–348, 1938.

[CAV 10] Cavarretta I., Coop M., O'Sullivan C., "The influence of particle characteristics on the behaviour of coarse grained soils", *Géotechnique*, vol. 60, no. 6, pp. 413–423, 2010.

[CAV 11] Cavarretta I., Rocchi I., Coop M.R., "A new interparticle friction apparatus for granular materials", *Canadian Geotechnical Journal*, vol. 48, no. 12, pp. 1829–1840, 2011.

[CHA 91] Chang C.S., Ma L., "A micromechanical-based micropolar theory for deformation of granular solids", *International Journal of Solids and Structures*, vol. 28, no. 1, pp. 67–86, 1991.

[CHA 06] Chang C.S., Kuhn M.R., "On virtual work and stress in granular media", *International Journal of Solids and Structures*, vol. 42, no. 13, pp. 6026–6051, 2006.

[CHE 90] Chen Y.-C., "Effect of inter-particle friction and initial fabric on fabric evolution", *Journal of the Chinese Institute of Engineers*, vol. 13, no. 2, pp. 147–156, 1990.

[CHE 08] Cheung G., O'Sullivan C., "Effective simulation of flexible lateral boundaries in two- and three-dimensional DEM simulations", *Particuology*, vol. 6, no. 6, pp. 483–500, 2008.

[CHO 06] Cho G.-C., Dodds J., Santamarina J.C., "Particle shape effects on packing density, stiffness, and strength: natural and crushed sands", *Journal of Geotechnical and Geoenvironmental Engineering*, vol. 132, no. 5, pp. 591–602, 2006.

[CHR 81] Christoffersen J., Mehrabadi M.M., Nemat-Nasser S., "A micromechanical description of granular material behavior", *Journal of Applied Mechanics*, vol. 48, no. 2, pp. 339–344, 1981.

[COL 07] Cole D.M., Peters J.F., "A physically based approach to granular media mechanics: grain-scale experiments, initial results and implications to numerical modeling", *Granular Matter*, vol. 9, no. 5, pp. 309–321, 2007.

[COL 08] Cole D.M., Peters J.F., "Grain-scale mechanics of geologic materials and lunar simulants under normal loading", *Granular Matter*, vol. 10, no. 3, pp. 171–185, 2008.

[COL 10] COLE D.M., MATHISEN L.U., HOPKINS M.A. *et al.*, "Normal and sliding contact experiments on gneiss", *Granular Matter*, vol. 12, pp. 69–86, 2010.

[CRE 04] CRESSWELL A., POWRIE W., "Triaxial tests on an unbonded locked sand", *Géotechnique*, vol. 54, no. 2, pp. 107–115, 2004.

[CUN 79] CUNDALL P.A., STRACK O.D.L., "A discrete numerical model for granular assemblies", *Géotechnique*, vol. 29, no. 1, pp. 47–65, 1979.

[CUN 82] CUNDALL P.A., DRESCHER A., STRACK O.D.L., "Numerical experiments on granular assemblies: measurements and observations", in VERMEER P., LUGER H. (eds.), *Deformation and Failure of Granular Materials*, A.A. Balkema, Rotterdam, 1982.

[CUN 83] CUNDALL P.A., STRACK O.D.L., "Modeling of microscopic mechanisms in granular material", in JENKINS J., SATAKE M. (eds.), *Mechanics of Granular Materials: New Models and Constitutive Relations*, Elsevier Science Pub. B.V., Amsterdam, 1983.

[CUN 88a] CUNDALL P.A., "Computer simulations of dense sphere assemblies", in SATAKE M., JENKINS J. (eds.), *Micromechanics of Granular Materials*, Elsevier Science Pub. B.V., Amsterdam, 1988.

[CUN 88b] CUNDALL P., "Formulation of a three-dimensional distinct element model–Part I. A scheme to detect and represent contacts in a system composed of many polyhedral blocks", *International Journal of Rock Mechanics and Mining Sciences & Geomechanics Abstracts*, vol. 25, no. 3, pp. 107–116, 1988.

[CUN 89a] CUNDALL P., "Numerical modeling of discontinua", in MUSTOE G.G.W., HENRIKSEN M., HUTTELMAIER H.-P. (eds.), *1st US Conference on Discrete Elements*, CSM Press, Golden, 1989.

[CUN 89b] CUNDALL P., "Numerical experiments on localization in frictional materials", *Ingenieur-Archiv*, vol. 59, no. 2, pp. 148–159, 1989.

[DAC 05] DA CRUZ F., EMAM S., PROCHNOW M. *et al.*, "Rheophysics of dense granular materials: discrete simulation of plane shear flows", *Physical Review E*, vol. 72, no. 2, p. 021309, 2005.

[DAN 57] DANTU P., "Contribution à l'Étude Mécanique et Géométrique des Milieux Pulvérulents", *Proceedings of the 4th International Conference on Soil Mechanics and Foundation Engineering*, vol. 1, pp. 144–148, 1957.

[DAO 13] DAOUADJI A., HICHER P.Y., JRAD M. *et al.*, "Experimental and numerical investigation of diffuse instability in granular materials using a microstructural model under various loading paths", *Géotechnique*, vol. 63, no. 5, p. 368, 2013.

[DEH 72] DEHOFF R.T., AIGELTINGER E.H., CRAIG K.R., "Experimental determination of the topological properties of three-dimensional microstructures", *Journal of Microscopy*, vol. 95, no. 1, pp. 69–91, 1972.

[DES 85] DESRUES J., LANIER J., STUTZ P., "Localization of the deformation in tests on sand sample", *Engineering Fracture Mechanics*, vol. 21, no. 4, pp. 909–921, 1985.

[DES 90a] DESRUES J., "Shear band initiation in granular materials: experimentation and theory", in DARVE F. (ed.), *Geomaterials: Constitutive Equations and Modelling*, Elsevier, London, 1990.

[DES 96] Desrues J., Chambon R., Mokni M. *et al.*, "Void ratio evolution inside shear bands in triaxial sand specimens studied by computed tomography", *Géotechnique*, vol. 46, no. 3, pp. 529–546, 1996.

[DES 04] Desrues J., Viggiani G., "Strain localization in sand: an overview of the experimental results obtained in Grenoble using stereophotogrammetry", *International Journal for Numerical and Analytical Methods in Geomechanics*, vol. 28, pp. 279–321, 2004.

[DID 01] Didwania A.K., Ledniczky K., Goddard J.D., "Kinematic diffusion in quasi-static granular deformation", *Quarterly Journal of Mechanics and Applied Mathematics*, vol. 54, no. 3, pp. 413–429, 2001.

[DIN 07] Ding W., Howard A.J., Peri M.D.M. *et al.*, "Rolling resistance moment of microspheres on surfaces: contact measurements", *Philosophical Magazine*, vol. 87, no. 36, pp. 5685–5696, 2007.

[DOB 91] Dobry R., Ng T.-T., Petrakis E. *et al.*, "General model for contact law between two rough spheres", *Journal of Engineering Mechanics*, vol. 117, no. 6, pp. 1365–1381, 1991.

[DOM 95] Dominik C., Tielens A.G.G.M., "Resistance to rolling in the adhesive contact of two elastic spheres", *Philosophical Magazine A*, vol. 72, no. 3, pp. 783–803, 1995.

[DOO 04] Doolin D.M., Sitar N., "Time integration in discontinuous deformation analysis", *Journal of Engineering Mechanics*, vol. 130, no. 3, pp. 249–258, 2004.

[DOW 79] Dowson D., *History of Tribology*, Longman, London, 1979.

[DRE 72] Drescher A., de Josselin de Jong G., "Photoelastic verification of a mechanical model for the flow of a granular material", *Journal of the Mechanics and Physics of Solids*, vol. 20, pp. 337–351, 1972.

[DUF 57] Duffy J., Mindlin R.D., "Stressstrain relations and vibrations of a granular media", *Journal of Applied Mechanics*, vol. 24, pp. 585–593, 1957.

[DUR 10] Durán O., Kruyt N.P., Luding S., "Micro-mechanical analysis of deformation characteristics of three-dimensional granular materials", *International Journal of Solids and Structures*, vol. 47, no. 17, pp. 2234–2245, 2010.

[ERI 67] Eringen A.C., *Mechanics of Continua*, John Wiley and Sons, New York, 1967.

[EST 08] Estrada N., Taboada A., Radjai F., "Shear strength and force transmission in granular media with rolling resistance", *Physical Review E*, vol. 78, no. 2, p. 021301, 2008.

[EST 11] Estrada N., Azéma E., Radjai F. *et al.*, "Identification of rolling resistance as a shape parameter in sheared granular media", *Physical Review E*, vol. 84, no. 1, p. 011306, 2011.

[FAV 01] Favier J.F., Abbaspour-Fard M.H., Kremmer M., "Modeling nonspherical particles using multisphere discrete elements", *Journal of Engineering Mechanics*, vol. 127, no. 10, pp. 971–977, 2001.

[FAZ 06] Fazekas S., Török J., Kertész J. *et al.*, "Morphologies of three-dimensional shear bands in granular media", *Physical Review E*, vol. 74, no. 3, p. 031303, 2006.

[FEN 01] Feng Y.T., Han K., Owen D.R.J., "Filling domains with disks", *Proceedings ICADDD-4, 4th International Conference on Analysis of Discontinuous Deformations*, Glasgow, pp. 239–250, 2001.

[FER 10] FERELLEC J.-F., McDOWELL G.R., "A method to model realistic particle shape and inertia in DEM", *Granular Matter*, vol. 12, no. 5, pp. 459–467, 2010.

[FIN 97a] FINNO R.J., ALARCON M.A., MOONEY M.A. *et al.*, "Shear bands in plane strain active tests of moist tamped and pluviated sands", *Proceedings of the 14th International Conference on Soil Mechanics and Foundation Engineering*, vol. 1, pp. 295–298, A.A. Balkema, Rotterdam, 1997.

[FIN 97b] FINNO R.J., HARRIS W.W., MOONEY M.A. *et al.*, "Shear bands in plane strain compression of loose sand", *Géotechnique*, vol. 47, no. 1, pp. 149–165, 1997.

[FON 13] FONSECA J., O'SULLIVAN C., COOP M.R. *et al.*, "Quantifying the evolution of soil fabric during shearing using scalar parameters", *Géotechnique*, vol. 63, no. 10, pp. 818–829, 2013.

[FUC 14] FUCHS R., WEINHART T., MEYER J. *et al.*, "Rolling, sliding and torsion of micron-sized silica particles: experimental, numerical and theoretical analysis", *Granular Matter*, vol. 16, no. 3, pp. 281–297, 2014.

[GAO 04] GAO J., LUEDTKE W.D., GOURDON D. *et al.*, "Frictional forces and Amontons' law: from the molecular to the macroscopic scale", *Journal of Physical Chemistry B*, vol. 108, no. 11, pp. 3410–3425, 2004.

[GER 73] GERMAIN P., "The method of virtual power in continuum mechanics. Part 2: microstructure", *SIAM Journal on Applied Mathematics*, vol. 25, no. 3, pp. 556–575, 1973.

[GOD 90] GODDARD J.D., "Nonlinear elasticity and pressure-dependent wave speeds in granular media", *Proceedings of the Royal Society of London A*, vol. 430, pp. 105–131, 1990.

[GON 12] GONG G., THORNTON C., CHAN A.C., "DEM simulations of undrained triaxial behavior of granular material", *Journal of Engineering Mechanics*, vol. 138, no. 6, pp. 560–566, 2012.

[GOO 62] GOODMAN L.E., BROWN C.B., "Energy dissipation in contact friction: constant normal and cyclic tangential loading", *Journal of Applied Mechanics*, vol. 29, no. 1, pp. 17–22, 1962.

[GRE 66] GREENWOOD J.A., WILLIAMSON J.B.P., "Contact of nominally flat surfaces", *Proceedings of the Royal Society of London A*, vol. 295, no. 1442, pp. 300–319, 1966.

[GRE 67] GREENWOOD J.A., TRIPP J.H., "The elastic contact of rough spheres", *Journal of Applied Mechanics*, vol. 34, no. 1, pp. 153–159, 1967.

[GRI 97] GRIGORIEV I.S., MEILIKHOV E.Z., *Handbook of Physical Quantities*, CRC Press, Boca Raton, 1997.

[GUO 13] GUO N., ZHAO J., "The signature of shear-induced anisotropy in granular media", *Computers and Geotechnics*, vol. 47, pp. 1–15, 2013.

[GUO 14] GUO N., ZHAO J., "Local fluctuations and spatial correlations in granular flows under constant-volume quasistatic shear", *Physical Review E*, vol. 89, p. 042208, 2014.

[HAL 10] HALL S.A., MUIRWOOD D., IBRAIM E. *et al.*, "Localised deformation patterning in 2D granular materials revealed by digital image correlation", *Granular Material*, vol. 12, no. 1, pp. 1–14, 2010.

[HAM 63] HAMILTON G.M., "Plastic flow in rollers loaded above the yield point", *Proceedings of the Institution of Mechanical Engineers*, vol. 177, no. 1, pp. 667–675, 1963.

[HAN 91] HAN C., VARDOULAKIS I.G., "Plane-strain compression experiments on water-saturated fine-grained sand", *Géotechnique*, vol. 41, no. 1, pp. 49–78, 1991.

[HAN 93] HAN C., DRESCHER A., "Shear bands in biaxial tests on dry coarse sand", *Soils and Foundations*, vol. 33, no. 1, pp. 118–132, 1993.

[HAR 95] HARRIS W.W., VIGGIANI G., MOONEY M.A. *et al.*, "Use of sterophotogrammetry to annalyze the development of shear bands in sand", *Geotechnical Testing Journal*, vol. 18, no. 4, pp. 405–420, 1995.

[HE 14] HE H., QIAN L., PANTANO C.G. *et al.*, "Mechanochemical wear of soda lime silica glass in humid environments", *Journal of the American Ceramic Society*, vol. 97, no. 7, pp. 2061–2068, 2014.

[HEC 97] HECKER C., "Physics, part 4: the third dimension", *Game Developer*, pp. 15–26, June 1997.

[HER 82] HERTZ H., "Über die berührung fester elastischer körper (on the contact of elastic solids)", *Journal of Reine Angewandte Mathematik*, vol. 92, pp. 156–171, 1882.

[HIL 62] HILL R., "Acceleration waves in solids", *Journal of the Mechanics and Physics of Solids*, vol. 10, no. 1, pp. 1–16, 1962.

[HIL 03] HILPERT M., GLANTZ R., MILLER C.T., "Calibration of a pore-network model by a pore-morphological analysis", *Transport in Porous Media*, vol. 51, no. 3, pp. 267–285, 2003.

[HOP 04] HOPKINS M.A., "Discrete element modeling with dilated particles", *Engineering with Computers*, vol. 21, nos. 2/3/4, pp. 422–430, 2004.

[HOR 62] HORN H.M., DEERE D.U., "Frictional characteristics of minerals", *Géotechnique*, vol. 12, no. 4, pp. 319–335, 1962.

[HOS 12] HOSSEININIA E.S., "Investigating the micromechanical evolutions within inherently anisotropic granular materials using discrete element method", *Granular Material*, vol. 14, no. 4, pp. 483–503, 2012.

[HUA 13] HUANG J., DA SILVA M.V., KRABBENHOFT K., "Three-dimensional granular contact dynamics with rolling resistance", *Computers and Geotechnics*, vol. 49, pp. 289–298, 2013.

[HUA 14] HUANG X., HANLEY K.J., O'SULLIVAN C. *et al.*, "Exploring the influence of interparticle friction on critical state behaviour using DEM", *International Journal for Numerical and Analytical Methods in Geomechanics*, vol. 38, no. 12, pp. 1276–1297, 2014.

[ITA 14] ITASCA, PFC – Particle Flow Code, Ver. 5.0, Itasca Consulting Group, Inc., Minneapolis, 2014.

[IWA 98] IWASHITA K., ODA M., "Rolling resistance at contacts in simulation of shear band development by DEM", *Journal of Engineering Mechanics*, vol. 124, no. 3, pp. 285–292, 1998.

[JÄG 99] JÄGER J., "Uniaxial deformation of a random packing of particles", *Archive of Applied Mechanics*, vol. 69, no. 3, pp. 181–203, 1999.

[JÄG 05] JÄGER J., *New Solutions in Contact Mechanics*, WIT Press, Southampton, 2005.

[JEA 95] JEAN M., "Frictional contact in collections of rigid or deformable bodies: numerical simulation of geomaterial motions", in SELVADURAI A.A.S., BOULON M.J. (eds.), *Mechanics of Geomaterial Interfaces*, Elsevier Science, Amsterdam, 1995.

[JEA 99] JEAN M., "The non-smooth contact dynamics method", *Computer Methods in Applied Mechanics and Engineering*, vol. 177, no. 3, pp. 235–257, 1999.

[JIA 05] JIANG J.J., YU H.-S., HARRIS D., "A novel discrete model for granular material incorporating rolling resistance", *Computers and Geotechnics*, vol. 32, pp. 340–357, 2005.

[JOH 55] JOHNSON K.L., "Surface interaction between elastically loaded bodies under tangential forces", *Proceedings of the Royal Society of London A*, vol. 230, no. 1183, pp. 531–548, 1955.

[JOH 85] JOHNSON K.L., *Contact Mechanics*, Cambridge University Press, 1985.

[KAL 67] KALKER J.J., On the Rolling Contact of Two Elastic Bodies in the Presence of Dry Friction, PhD Thesis, Delft University of Technology, 1967.

[KAL 00] KALKER J.J., "Rolling contact phenomena – linear elasticity", in JACOBSON B., KALKER J.J. (eds.), *Rolling Contact Phenomena*, no. 411, Courses and Lectures, CISM, Springer, Vienna, 2000.

[KAN 79] KANATANI K.-I., "A micropolar continuum theory for the flow of granular materials", *International Journal of Engineering Science*, vol. 17, no. 4, pp. 419–432, 1979.

[KAN 84] KANATANI K.-I., "Distribution of directional data and fabric tensors", *International Journal of Engineering Science*, vol. 22, no. 2, pp. 149–164, 1984.

[KE 95] KE T.-C., BRAY J., "Modeling of particulate media using discontinuous deformation analysis", *Journal of Engineering Mechanics*, vol. 121, no. 11, pp. 1234–1243, 1995.

[KIS 88] KISHINO Y., "Disc model analysis of granular media", in SATAKE M., JENKINS J.T. (eds.), *Micromechanics of Granular Materials*, Elsevier Science Publication, 1988.

[KIS 03] KISHINO Y., "On the incremental nonlinearity observed in a numerical model for granular media", *Revista Italiana Di Geotecnica*, vol. 3, pp. 30–38, 2003.

[KNU 73] KNUTH D.E., *The Art of Computer Programming: Fundamental Algorithms*, vol. 1, Addison-Wesley Publication, Reading, 1973.

[KON 82] KONISHI J., ODA M., NEMAT-NASSER S., "Inherent anisotropy and shear strength of assembly of oval cross-sectional rods", in VERMEER P., LUGER H. (eds.), *Deformation and Failure of Granular Materials*, A.A. Balkema Publication, Rotterdam, 1982.

[KON 88] KONISHI J., NARUSE F., "A note on fabric in terms of voids", in SATAKE M., JENKINS J. (eds.), *Micromechanics of Granular Materials*, Elsevier Science Pub. B.V., Amsterdam, 1988.

[KRE 01] KREMMER M., FAVIER J.F., "A method for representing boundaries in discrete element modelling – part I: geometry and contact detection", *International Journal for Numerical Methods in Engineering*, vol. 51, no. 12, pp. 1407–1421, 2001.

[KRU 63] KRUMBEIN W.C., SLOSS L.L., *Stratigraphy and Sedimentation*, W.H. Freeman and Company, San Francisco, 1963.

[KRU 96] Kruyt N.P., Rothenburg L., "Micromechanical definition of the strain tensor for granular materials", *Journal of Applied Mechanics*, vol. 118, no. 3, pp. 706–711, 1996.

[KRU 03] Kruyt N.P., "Statics and kinematics of discrete Cosserat-type granular materials", *International Journal of Solids and Structures*, vol. 40, no. 3, pp. 511–534, 2003.

[KRU 06] Kruyt N.P., Rothenburg L., "Shear strength, dilatancy, energy and dissipation in quasi-static deformation of granular materials", *Journal of Statistical Mechanics: Theory and Experiment*, vol. 2006, no. 07, p. P07021, 2006.

[KRU 07] Kruyt N.P., Antony S.J., "Force, relative-displacement, and work networks in granular materials subjected to quasistatic deformation", *Physical Review E*, vol. 75, no. 5, p. 051308, 2007.

[KRU 09] Kruyt N.P., Rothenburg L., "Plasticity of granular materials: a structural-mechanics view", *AIP Conference Proceedings*, vol. 1145, p. 1073, 2009.

[KRU 10] Kruyt N.P., "Micromechanical study of plasticity of granular materials", *Comptes Rendus Mécanique*, vol. 338, no. 10, pp. 596–603, 2010.

[KRU 12] Kruyt N.P., "Micromechanical study of fabric evolution in quasi-static deformation of granular materials", *Mechanics of Materials*, vol. 44, pp. 120–129, 2012.

[KRU 14] Kruyt N.P., Rothenburg L., "On micromechanical characteristics of the critical state of two-dimensional granular materials", *Acta Mechanica*, vol. 225, no. 8, pp. 2301–2318, Springer Vienna, 2014.

[KUH 91] Kuhn M.R., "Factors affecting the incremental stiffness of particle assemblies", in Adeli H., Sierakowski R.L. (eds.), *Mechanics Computing in 1990's and Beyond*, vol. 2, ASCE, New York, 1991.

[KUH 92] Kuhn M.R., Mitchell J.K., "Modelling of soil creep with the discrete element method", *Engineering with Computers*, vol. 9, no. 2, pp. 277–287, April 1992.

[KUH 95] Kuhn M.R., "A flexible boundary for three-dimensional DEM particle assemblies", *Engineering with Computers*, vol. 12, no. 2, pp. 175–183, 1995.

[KUH 97] Kuhn M.R., "Deformation measures for granular materials", in Chang C.S., Misra A., Liang R.Y. *et al.*, (eds.), *Mechanics of Deformation and Flow of Particulate Materials*, ASCE, New York, 1997.

[KUH 99] Kuhn M.R., "Structured deformation in granular materials", *Mechanics of Materials*, vol. 31, no. 6, pp. 407–429, 1999.

[KUH 02] Kuhn M.R., Bagi K., "Generalized continuum models for granular materials: bridging a transition from micro to macro", in Pyrz R., Schjødt-Thomsen J., Rauhe J.C. *et al.*,(eds.), *Proceedings of International Conference on New Challenges in Mesomechanics*, vol. 1, Aalborg, pp. 105–111, 2002.

[KUH 03a] Kuhn M.R., "Heterogeneity and patterning in the quasi-static behavior of granular materials", *Granular Material*, vol. 4, no. 4, pp. 155–166, 2003.

[KUH 03b] Kuhn M.R., "Smooth convex three-dimensional particle for the discrete element method", *Journal of Engineering Mechanics*, vol. 129, no. 5, pp. 539–547, 2003.

[KUH 04a] KUHN M.R., "Boundary integral for gradient averaging in two dimensions: application to polygonal regions in granular materials", *International Journal for Numerical Methods in Engineering*, vol. 59, no. 4, pp. 559–576, 2004.

[KUH 04b] KUHN M.R., "Rates of stress in dense unbonded frictional materials during slow loading", in ANTONY S.J., HOYLE W., DING Y. (eds.), *Advances in Granular Materials: Fundamentals and Applications*, Royal Society of Chemistry, London, 2004.

[KUH 04c] KUHN M.R., BAGI K., "Alternative definition of particle rolling in a granular assembly", *Journal of Engineering Mechanics*, vol. 130, no. 7, pp. 826–835, 2004.

[KUH 04d] KUHN M.R., BAGI K., "Contact rolling and deformation in granular media", *International Journal of Solids and Structures*, vol. 41, no. 21, pp. 5793–5820, 2004.

[KUH 05a] KUHN M.R., "Are granular materials simple? An experimental study of strain gradient effects and localization.", *Mechanics of Materials*, vol. 37, no. 5, pp. 607–627, 2005.

[KUH 05b] KUHN M.R., BAGI K., "On the relative motions of two rigid bodies at a compliant contact: application to granular media.", *Mechanics Research Communications*, vol. 32, no. 4, pp. 463–480, 2005.

[KUH 06] KUHN M.R., CHANG C.S., "Stability, bifurcation, and softening in discrete systems: a conceptual approach for granular materials", *International Journal of Solids and Structures*, vol. 43, no. 20, pp. 6026–6051, 2006.

[KUH 09] KUHN M.R., BAGI K., "Specimen size effect in discrete element simulations of granular assemblies", *Journal of Engineering Mechanics*, vol. 135, no. 6, pp. 485–492, 2009.

[KUH 10] KUHN M.R., "Micro-mechanics of fabric and failure in granular materials", *Mechanics of Materials*, vol. 42, no. 9, pp. 827–840, 2010.

[KUH 11] KUHN M.R., "Implementation of the Jäger contact model for discrete element simulations", *International Journal for Numerical Methods in Engineering*, vol. 88, no. 1, pp. 66–82, 2011.

[KUH 14a] KUHN M.R., "Dense granular flow at the critical state: maximum entropy and topological disorder", *Granular Material*, vol. 16, no. 4, pp. 499–508, 2014.

[KUH 14b] KUHN M.R., "Transient rolling friction model for discrete element simulations of sphere assemblies", *Comptes Rendus Mécanique*, vol. 342, no. 3, pp. 129–140, 2014.

[KUH 14c] KUHN M.R., RENKEN H., MIXSELL A. *et al.*, "Investigation of cyclic liquefaction with discrete element simulations", *Journal of Geotechnical and Geoenvironmental Engineering*, vol. 140, no. 12, p. 04014075, 2014.

[KUH 15] KUHN M.R., SUN W., WANG Q., "Stress-induced anisotropy in granular materials: fabric, stiffness, and permeability", *Acta Geotechnica*, vol. 10, no. 4, pp. 399–419, 2015.

[KUH 16a] KUHN M.R., "Contact transience during slow loading of dense granular materials", *Journal of Engineering Mechanics*, vol. 143, no. 1, p. C4015003, 2016.

[KUH 16b] KUHN M.R., "The critical state of granular media: convergence, stationarity and disorder", *Géotechnique*, vol. 66, no. 11, pp. 902–909, 2016.

[KUH 16c] KUHN M.R., "Maximum disorder model for dense steady-state flow of granular materials", *Mechanics of Materials*, vol. 93, pp. 63–80, 2016.

[KUO 98] KUO C.-Y., FROST J.D., CHAMEAU J.-L.A., "Image analysis determination of steriology based fabric tensors", *Géotechnique*, vol. 48, no. 4, pp. 515–525, 1998.

[KWI 90] KWIECIEN M.J., MACDONALD I.F., DULLIEN F.A.L., "Three-dimensional reconstruction of porous media from serial section data", *Journal of Microscopy*, vol. 159, no. 3, pp. 343–359, 1990.

[LAM 69] LAMBE T.W., WHITMAN R.V., *Soil Mechanics*, John Wiley & Sons, New York, 1969.

[LAN 01] LANIER J., "Micro-mechanisms of deformation in granular materials: experiments and numerical results", in VERMEER P.A., DIEBELS S., EHLERS W. *et al.*(eds.), *Continuous and Discontinuous Modelling of Cohesive-Frictional Materials*, Springer, Berlin, 2001.

[LI 02a] LI S., LIU W.K., "Meshfree and particle methods and their applications", *Applied Mechanics Reviews*, vol. 55, no. 1, pp. 1–34, 2002.

[LI 02b] LI X., DAFALIAS Y., "Constitutive modeling of inherently anisotropic sand behavior", *Journal of Geotechnical and Geoenvironmental Engineering*, vol. 128, no. 10, pp. 868–880, 2002.

[LI 09] LI X., LI X.-S., "Micro-macro quantification of the internal structure of granular materials", *Journal of Engineering Mechanics*, vol. 135, no. 7, pp. 641–656, 2009.

[LIA 97] LIAO C.-L., CHANG T.-P., YOUNG D.-H. *et al.*, "Stress-strain relationship for granular materials based on the hypothesis of best fit", *International Journal of Solids and Structures*, vol. 34, no. 31–32, pp. 4087–4100, 1997.

[LIA 00] LIANG Z., IOANNIDIS M.A., CHATZIS I., "Geometric and topological analysis of three-dimensional porous media: pore space partitioning based on morphological skeletonization", *Journal of Colloid and Interface Science*, vol. 221, no. 1, pp. 13–24, 2000.

[LIM 14] LIM K.-W., ANDRADE J.E., "Granular element method for three-dimensional discrete element calculations", *International Journal for Numerical and Analytical Methods in Geomechanics*, vol. 38, no. 2, pp. 167–188, 2014.

[LIN 96] LINDQUIST W.B., LEE S.-M., COKER D.A. *et al.*, "Medial axis analysis of void structure in three-dimensional tomographic images of porous media", *Journal of Geophysical Research: Solid Earth*, vol. 101, no. B4, pp. 8297–8310, 1996.

[LIN 97] LIN X., NG T.-T., "A three-dimensional discrete element model using arrays of ellipsoids", *Géotechnique*, vol. 47, no. 2, pp. 319–329, 1997.

[MAG 08] MAGOARIEC H., DANESCU A., CAMBOU B., "Nonlocal orientational distribution of contact forces in granular samples containing elongated particles", *Acta Geotechnica*, vol. 3, no. 1, pp. 49–60, 2008.

[MAJ 05] MAJMUDAR T.S., BEHRINGER R.P., "Contact force measurements and stress-induced anisotropy in granular materials", *Nature*, vol. 435, no. 1079, pp. 1079–1082, 2005.

[MAL 69] MALVERN L.E., *Introduction to the Mechanics of a Continuous Medium*, Prentice Hall, Englewood Cliffs, 1969.

[MAN 62] MANDEL J., "Ondes plastiques dans un milieu indéfini à trois dimensions", *Journal of Mécanique*, vol. 1, no. 1, pp. 3–30, 1962.

[MAT 02] MATSUSHIMA T., ISHII T., KONAGAI K., "Observation of grain motion in the interior of a PSC test specimen by laser-aided tomography", *Soils and Foundations*, vol. 42, no. 5, pp. 27–36, 2002.

[MAT 03] MATSUSHIMA T., SAOMOTO H., TSUBOKAWA Y. *et al.*, "Observation of grain rotation inside granular assembly during shear deformation", *Soils and Foundations*, vol. 43, no. 4, pp. 95–106, 2003.

[MIC 78] MICHAŁOWSKI R., MRÓZ Z., "Associated and non-associated sliding rules in contact friction problems", *Archives of Mechanics: Archiwum Mechaniki Stosowanej*, vol. 30, no. 3, pp. 259–276, 1978.

[MIC 01] MICHIELSEN K., DE RAEDT H., "Integral-geometry morphological image analysis", *Physics Reports*, vol. 347, no. 6, pp. 461–538, 2001.

[MIN 49] MINDLIN R.D., "Compliance of elastic bodies in contact", *Journal of Applied Mechanics*, vol. 16, pp. 259–268, 1949.

[MIN 53] MINDLIN R., DERESIEWICZ H., "Elastic spheres in contact under varying oblique forces", *Journal of Applied Mechanics*, vol. 19, no. 1, pp. 327–344, 1953.

[MIS 98] MISRA A., "Biaxial shear of granular materials", in MURAKAMI H., LUCO J.E. (eds.), *Engineering Mechanics: A Force for the 21st Century*, ASCE, Reston, 1998.

[MIT 05] MITCHELL J.K., SOGA K., *Fundamentals of Soil Behavior*, 3rd edition, John Wiley & Sons, 2005.

[MOH 10] MOHAMED A., GUTIERREZ M., "Comprehensive study of the effects of rolling resistance on the stress–strain and strain localization behavior of granular materials", *Granular Matter*, vol. 12, no. 5, pp. 527–541, Springer, 2010.

[MON 88] MONTANA D.J., "The kinematics of contact and grasp", *The International Journal of Robotics Research*, vol. 7, no. 3, pp. 17–32, 1988.

[MOR 94] MOREAU J.J., "Some numerical methods in multibody dynamics: application to granular materials", *European Journal of Mechanics ? A/Solids*, vol. 13, pp. 93–114, 1994.

[MOR 04] MOREAU J.J., "An introduction to unilateral dynamics", in FRÉMOND M., MACERI F. (eds.), *Lecture Notes in Applied and Computational Mechanics*, vol. 14, Springer-Verlag, 2004.

[MUT 07] MUTH B., MÜLLER M.-K., EBERHARD P. *et al.*, "Collision detection and administration methods for many particles with different sizes", available at: http://doc.utwente.nl/80337/1/Muth07collision.pdf, 2007.

[NG 04] NG T.-T., "Triaxial test simulations with discrete element method and hydrostatic boundaries", *Journal of Engineering Mechanics*, vol. 130, no. 10, pp. 1188–1194, 2004.

[NG 06] NG T.-T., "Input parameters of discrete element methods", *Journal of Engineering Mechanics*, vol. 132, no. 7, pp. 723–729, 2006.

[NGU 09] Nguyen N.-S., Magoariec H., Cambou B. et al., "Analysis of structure and strain at the meso-scale in 2D granular materials", International Journal of Solids and Structures, vol. 46, no. 17, pp. 3257–3271, 2009.

[NOU 03] Nouguier-Lehon C., Cambou B., Vincens E., "Influence of particle shape and angularity on the behaviour of granular materials: a numerical analysis", International Journal for Numerical and Analytical Methods in Geomechanics, vol. 27, no. 14, pp. 1207–1226, 2003.

[NOU 05] Nouguier-Lehon C., Vincens E., Cambou B., "Structural changes in granular materials: the case of irregular polygonal particles", International Journal of Solids and Structures, vol. 42, no. 24–25, pp. 6356–6375, 2005.

[NOU 10] Nouguier-Lehon C., "Effect of the grain elongation on the behaviour of granular materials in biaxial compression", Comptes Rendus Mécanique, vol. 338, no. 10, pp. 587–595, 2010.

[NOV 94] Nova R., "Controllability of the incremental response of soil specimens subjected to arbitrary loading programmes", Journal of the Mechanical Behavior of Materials, vol. 5, no. 2, pp. 193–201, 1994.

[ODA 72a] Oda M., "Initial fabrics and their relations to mechanical properties of granular material", Soils and Foundations, vol. 12, no. 1, pp. 17–36, 1972.

[ODA 72b] Oda M., "The mechanism of fabric changes during compressional deformation of sand", Soils and Foundations, vol. 12, no. 2, pp. 1–18, 1972.

[ODA 77] Oda M., "Co-ordination number and its relation to shear strength of granular material", Soils and Foundations, vol. 17, no. 2, pp. 29–42, 1977.

[ODA 82] Oda M., Konishi J., Nemat-Nasser S., "Experimental micromechanical evaluation of strength of granular materials: effects of particle rolling", Mechanics of Materials, vol. 1, no. 4, pp. 269–283, 1982.

[ODA 85] Oda M., Nemat-Nasser S., Konishi J., "Stress-induced anisotropy in granular masses", Soils and Foundations, vol. 25, no. 3, pp. 85–97, 1985.

[ODA 97] Oda M., Iwashita K., Kakiuchi T., "Importance of particle rotation in the mechanics of granular materials", in Behringer R.P., Jenkins J.T. (eds.), Powders & Grains 97, A. A. Balkema, Rotterdam, 1997.

[ODA 98a] Oda M., Iwashita K., "Couple stress developed in shear bands (1): particle rotation and couple stress in granular media", in Murakami H., Luco J.E. (eds.), Engineering Mechanics: A Force for the 21st Century, ASCE, Reston, 1998.

[ODA 98b] Oda M., Kazama H., "Microstructure of shear bands and its relation to the mechanisms of dilatancy and failure of dense granular soils", Géotechnique, vol. 48, no. 4, pp. 465–481, 1998.

[OGE 98] Oger L., Savage S.B., Corriveau D. et al., "Yield and deformation of an assembly of disks subjected to a deviatoric stress loading", Mechanics of Materials, vol. 27, no. 4, pp. 189–210, 1998.

[OSU 03] O'SULLIVAN C., BRAY J.D., LI S., "A new approach for calculating strain for particulate media", *International Journal for Numerical and Analytical Methods in Geomechanics*, vol. 27, no. 10, pp. 859–877, 2003.

[OSU 04] O'SULLIVAN C., BRAY J.D., "Selecting a suitable time step for discrete element simulations that use the central difference time integration scheme", *Engineering with Computers*, vol. 21, no. 2/3/4, pp. 278–303, 2004.

[OTS 15] OTSUBO M., O'SULLIVAN C., SIM W.W. *et al.*, "Quantitative assessment of the influence of surface roughness on soil stiffness", *Géotechnique*, vol. 65, no. 8, pp. 694–700, 2015.

[OUA 01] OUADFEL H., ROTHENBURG L., "Stress–force–fabric relationship for assemblies of ellipsoids", *Mechanics of Materials*, vol. 33, no. 4, pp. 201–221, 2001.

[PEÑ 07] PEÑA A.A., GARCÍA-ROJO R., HERRMANN H.J., "Influence of particle shape on sheared dense granular media", *Granular Matter*, vol. 9, pp. 279–291, 2007.

[PEÑ 09] PEÑA A.A., GARCÍA-ROJO R., ALONSO-MARROQUÍN F. *et al.*, "Investigation of the critical state in soil mechanics using DEM", in NAKAGAWA M., LUDING S. (eds.), *Powders and Grains 2009*, American Institute of Physics, 2009.

[PET 05] PETERS J.F., MUTHUSWAMY M., WIBOWO J. *et al.*, "Characterization of force chains in granular material", *Physical Review E*, vol. 72, no. 4, p. 041307, APS, 2005.

[PET 13] PETERS J.F., WALIZER L.E., "Patterned nonaffine motion in granular media", *Journal of Engineering Mechanics*, vol. 139, no. 10, pp. 1479–1490, 2013.

[POT 04] POTYONDY D.O., CUNDALL P.A., "A bonded-particle model for rock", *International Journal of Rock Mechanics and Mining Sciences*, vol. 41, no. 8, pp. 1329–1364, 2004.

[POW 53] POWERS M.C., "A new roundness scale for sedimentary particles", *Journal of Sedimentary Research*, vol. 23, no. 2, pp. 117–119, 1953.

[PRA 91] PRASAD P.B., JERNOT J.P., "Topological description of the densification of a granular medium", *Journal of Microscopy*, vol. 163, no. 2, pp. 211–220, 1991.

[PRE 85] PREPARATA F.P., SHAMOS M.I., *Computational Geometry: An Introduction*, Springer-Verlag, New York, 1985.

[RAD 96] RADJAI F., JEAN M., MOREAU J.-J. *et al.*, "Force distributions in dense two-dimensional granular systems", *Physical Review Letters*, vol. 77, no. 2, pp. 274–277, 1996.

[RAD 97] RADJAI F., WOLF D.E., ROUX S. *et al.*, "Force networks in dense granular media", in BEHRINGER R.P., JENKINS J.T. (eds.), *Powders & Grains 97*, A. A. Balkema, Rotterdam, 1997.

[RAD 98] RADJAI F., WOLF D.E., JEAN M. *et al.*, "Bimodal character of stress transmission in granular packings", *Physical Review Letters*, vol. 80, no. 1, pp. 61–64, 1998.

[RAD 02] RADJAI F., ROUX S., "Turbulentlike fluctuations in quasistatic flow of granular media", *Physical Review Letters*, vol. 89, no. 6, p. 064302, 2002.

[RAD 09] RADJAI F., RICHEFEU V., "Contact dynamics as a nonsmooth discrete element method", *Mechanics of Materials*, vol. 41, no. 6, pp. 715–728, 2009.

[RAD 12] Radjai F., Delenne J.-Y., Azéma E. *et al.*, "Fabric evolution and accessible geometrical states in granular materials", *Granular Matter*, vol. 14, no. 2, pp. 259–264, 2012.

[REC 06] Rechenmacher A.L., "Grain-scale processes governing shear band initiation and evolution in sands", *Journal of the Mechanics and Physics of Solids*, vol. 54, no. 1, pp. 22–45, 2006.

[REC 10] Rechenmacher A., Abedi S., Chupin O., "Evolution of force chains in shear bands in sands", *Géotechnique*, vol. 60, no. 5, pp. 343–351, 2010.

[REC 11] Rechenmacher A.L., Abedi S., "Length scales for nonaffine deformation in localized, granular shear", in Bonelli S., Dascalu C., Nicot F. (eds.), *Advances in Bifurcation and Degradation in Geomaterials*, Springer, 2011.

[REE 96] Reeves P.C., Celia M.A., "A functional relationship between capillary pressure, saturation, and interfacial area as revealed by a pore-scale network model", *Water Resources Research*, vol. 32, no. 8, pp. 2345–2358, 1996.

[RIC 76] Rice J.R., "The localization of plastic deformation", in Koiter W.T. ed., *Theoretical and Applied Mechanics*, North-Holland Publishing, Amsterdam, 1976.

[RIC 12] Richefeu V., Combe G., Viggiani G., "An experimental assessment of displacement fluctuations in a 2D granular material subjected to shear", *Géotechnique Letters*, vol. 2, no. July–September, pp. 113–118, 2012.

[ROS 70] Roscoe K.H., "The influence of strains in soil mechanics", *Géotechnique*, vol. 20, no. 2, pp. 129–170, 1970.

[ROT 89] Rothenburg L., Bathurst R., "Analytical study of induced anisotropy in idealized granular materials", *Géotechnique*, vol. 39, no. 4, pp. 601–614, 1989.

[ROT 91] Rothenburg L., Bathurst R., "Numerical simulation of idealized granular assemblies with plane elliptical particles", *Computers and Geotechnics*, vol. 11, no. 4, pp. 315–329, 1991.

[ROT 92] Rothenburg L., Bathurst R., "Micromechanical features of granular assemblies with planar elliptical particles", *Géotechnique*, vol. 42, no. 1, pp. 79–95, 1992.

[ROT 93] Rothenburg L., Bathurst R., "Influence of particle eccentricity on micromechanical behavior of granular materials", *Mechanics of Materials*, vol. 16, pp. 141–152, 1993.

[ROT 04] Rothenburg L., Kruyt N.P., "Critical state and evolution of coordination number in simulated granular materials", *International Journal of Solids and Structures*, vol. 31, no. 21, pp. 5763–5774, 2004.

[ROU 04] Rougier E., Munjiza A., John N.W.M., "Numerical comparison of some explicit time integration schemes used in DEM, FEM/DEM and molecular dynamics", *International Journal for Numerical Methods in Engineering*, vol. 61, no. 6, pp. 856–879, 2004.

[ROU 08] Rousé P.C., Fannin R.J., Shuttle D.A., "Influence of roundness on the void ratio and strength of uniform sand", *Géotechnique*, vol. 58, no. 3, pp. 227–231, 2008.

[ROU 02] Roux J.-N., Combe G., "Quasistatic rheology and the origins of strain", *Comptes Rendus Physique*, vol. 3, no. 2, pp. 131–140, 2002.

[ROU 03] Roux J.-N., Combe G., "On the meaning and microscopic origins of quasistatic deformation of granular materials", *Proceedings of the 16th ASCE Engineering Mechanics Conference*, vol. Paper No. 759, pp. 1–5, ASCE, 2003.

[ROZ 10] Rozmanov D., Kusalik P.G., "Robust rotational-velocity-Verlet integration methods", *Physical Review E*, vol. 81, no. 5, p. 056706, 2010.

[SAA 99] Saada A.S., Liang L., Figueroa J.L. *et al.*, "Bifurcation and shear band propagation in sands", *Géotechnique*, vol. 49, no. 3, pp. 367–385, 1999.

[SAI 12] Saint-Cyr B., Szarf K., Voivret C. *et al.*, "Particle shape dependence in 2D granular media", *Europhysics Letters*, vol. 98, no. 4, p. 44008, 2012.

[SAL 09] Salot C., Gotteland P., Villard P., "Influence of relative density on granular materials behavior: DEM simulations of triaxial tests", *Granular Matter*, vol. 11, no. 4, pp. 221–236, 2009.

[SAN 98] Santamarina J.C., Cascante G., "Effect of surface roughness on wave propagation parameters", *Géotechnique*, vol. 48, no. 1, pp. 129–136, 1998.

[SAN 01] Santamarina J.C., Klein K.A., Fam M.A., *Soils and Waves: Particulate Materials Behavior, Characterization and Process Monitoring*, Wiley, New York, 2001.

[SAT 82] Satake M., "Fabric tensor in granular materials", in Vermeer P.A., Luger H.J. (eds.), *Proc. IUTAM Symp. on Deformation and Failure of Granular Materials*, A.A. Balkema, Rotterdam, 1982.

[SAT 92] Satake M., "A discrete-mechanical approach to granular materials", *International Journal of Engineering Science*, vol. 30, no. 10, pp. 1525–1533, 1992.

[SAT 93] Satake M., "Discrete-mechanical approach to granular media", in Thornton C. (ed.), *Powders & Grains 93*, A.A. Balkema, Rotterdam, 1993.

[SCA 82] Scarpelli G., Wood D.M., "Experimental observations of sthear band patterns in direct shear tests", in Vermeer P., Luger H. (eds.), *Deformation and Failure of Granular Materials*, A.A. Balkema, Rotterdam, 1982.

[SCH 68] Schofield A.N., Wroth P., *Critical State Soil Mechanics*, McGraw-Hill, New York, 1968.

[SCH 11] Schröder-Turk G.E., Mickel W., Kapfer S.C. *et al.*, "Minkowski tensor shape analysis of cellular, granular and porous structures", *Advanced Materials*, vol. 23, nos. 22–23, pp. 2535–2553, 2011.

[SCH 13] Schröder-Turk G.E., Mickel W., Kapfer S. *et al.*, "Minkowski tensors of anisotropic spatial structure", *New Journal of Physics*, vol. 15, no. 8, 083028, 2013.

[SEN 13] Senetakis K., Coop M.R., Todisco M.C., "Tangential load–deflection behaviour at the contacts of soil particles", *Géotechnique Letters*, vol. 3, no. 2, pp. 59–66, 2013.

[SER 82] Serra J., *Image Analysis and Mathematical Morphology*, Academic Press, London, 1982.

[SER 84] Seridi A., Dobry R., An incremental elastic-plastic model for the force–displacement relation at the contact between two elastic spheres, Research Report, Rensselaer Polytechnic Institute, Troy, 1984.

[SHA 83] Shahinpoor M., "Frequency distribution of voids in randomly packed monogranular layers", in Jenkins J., Satake M. (eds.), *Mechanics of Granular Materials: New Models and Constitutive Relations*, Elsevier Science Pub., Amsterdam, 1983.

[SHI 88] Shi G.H., Discontinuous deformation analysis: a new numerical model for the statics and dynamics of block systems, PhD Thesis, University of California, Berkeley, 1988.

[SHI 93] Shi G., *Block System Modeling by Discontinuous Deformation Analysis*, Computational Mechanics Publications, Southampton, 1993.

[SHI 12] Shin H., Santamarina J.C., "Role of particle angularity on the mechanical behavior of granular mixtures", *Journal of Geotechnical and Geoenvironmental Engineering*, vol. 139, no. 2, pp. 353–355, 2012.

[SHO 03] Shodja H.M., Nezami E.G., "A micromechanical study of rolling and sliding contacts in assemblies of oval granules", *International Journal for Numerical and Analytical Methods in Geomechanics*, vol. 27, no. 5, pp. 403–424, 2003.

[SIL 02] Silbert L.E., Ertaş D., Grest G.S. *et al.*, "Geometry of frictionless and frictional sphere packings", *Physical Review E*, vol. 65, no. 3, p. 031304, 2002.

[STR 11] Strang G., *Introduction to Linear Algebra*, 4th edition, Wellesley-Cambridge Press, Wellesley, 2011.

[SUN 10] Sun Q., Jin F., Liu J. *et al.*, "Understanding force chains in dense granular materials", *International Journal of Modern Physics B*, vol. 24, no. 29, pp. 5743–5759, 2010.

[SUN 13] Sun W., Kuhn M.R., Rudnicki J.W., "A multiscale DEM-LBM analysis on permeability evolutions inside a dilatant shear band", *Acta Geotechnica*, vol. 8, no. 5, pp. 465–480, 2013.

[SUZ 14] Suzuki K., Kuhn M.R., "Uniqueness of discrete element simulations in monotonic biaxial shear tests", *International Journal of Geomechanics*, vol. 14, no. 5, p. 06014010, 2014.

[TAT 90] Tatsuoka F., Nakamura S., Huang C.-C. *et al.*, "Strength anisotropy and shear band direction in plane strain tests of sand", *Soils and Foundations*, vol. 30, no. 1, pp. 35–54, 1990.

[THO 99] Thomas P.A., Bray J.D., "Capturing nonspherical shape of granular media with disk clusters", *Journal of Geotechnical and Geoenvironmental Engineering*, vol. 125, no. 3, pp. 169–178, 1999.

[THO 61] Thomas T.Y., *Plastic Flow and Fracture in Solids*, vol. 2, Academic Press, New York, 1961.

[THO 86] Thornton C., Barnes D.J., "Computer simulated deformation of compact granular assemblies", *Acta Mechanica*, vol. 64, nos. 1–2, pp. 45–61, 1986.

[THO 98] Thornton C., Antony S.J., "Quasi-static deformation of particulate media", *Philosophical Transactions of the Royal Society A*, vol. 356, no. 1747, pp. 2763–2782, 1998.

[THO 00] Thornton C., "Numerical simulations of deviatoric shear deformation of granular media", *Géotechnique*, vol. 50, no. 1, pp. 43–53, 2000.

[THO 06] THORNTON C., ZHANG L., "A numerical examination of shear banding and simple shear non-coaxial flow rules", *Philosophical Magazine*, vol. 86, nos. 21–22, pp. 3425–3452, 2006.

[THO 10] THORNTON C., ZHANG L., "On the evolution of stress and microstructure during general 3D deviatoric straining of granular media", *Géotechnique*, vol. 60, no. 5, pp. 333–341, 2010.

[TIN 91] TING J.M., "Ellipse-based micromechanical model for angular granular materials", in ADELI H., SIERAKOWSKI R.L. (eds.), *Mechanics Computing in 1990's and Beyond*, vol. 2, ASCE, New York, 1991.

[TOR 02] TORDESILLAS A., WALSH D.C., "Incorporating rolling resistance and contact anisotropy in micromechanical models of granular media", *Powder Technology*, vol. 124, no. 1, pp. 106–111, 2002.

[TOR 07] TORDESILLAS A., "Force chain buckling, unjamming transitions and shear banding in dense granular assemblies", *Philosophical Magazine*, vol. 87, no. 32, pp. 4987–5016, 2007.

[TOR 08] TORDESILLAS A., MUTHUSWAMY M., WALSH S.D.C., "Mesoscale measures of nonaffine deformation in dense granular assemblies", *Journal of Engineering Mechanics*, vol. 134, no. 12, pp. 1095–1113, 2008.

[TOR 09] TORDESILLAS A., ZHANG J., BEHRINGER R., "Buckling force chains in dense granular assemblies: physical and numerical experiments", *Geomechanics & Geoengineering: An International Journal*, vol. 4, no. 1, pp. 3–16, 2009.

[TOR 15] TORDESILLAS A., PUCILOWSKI S., TOBIN S. *et al.*, "Shear bands as bottlenecks in force transmission", *Europhysics Letters*, vol. 110, no. 5, p. 58005, 2015.

[TSU 01] TSUCHIKURA T., SATAKE M., "A consideration on the statistical analysis of particle packing using loop tensors", in KISHINO Y. (ed.), *Powders and Grains 2001*, A.A. Balkema, Lisse, 2001.

[TZA 95] TZAFEROPOULOS M.A., "On the numerical modeling of convex particle assemblies with friction", *Computer Methods in Applied Mechanics and Engineering*, vol. 127, nos. 1–4, pp. 371–386, 1995.

[UNG 03] UNGER T., KERTÉSZ J., "The contact dynamics method for granular media", in GARRIDO P.L., MARRO J. (eds.), *Modeling of Complex Systems*, vol. 661 of *American Institute of Physics Conference Series*, April 2003.

[UTT 04] UTTER B., BEHRINGER R.P., "Self-diffusion in dense granular shear flows", *Physical Review E*, vol. 69, no. 3, p. 031308, 2004.

[VAR 98] VARDOULAKIS I., "Strain localization in granular materials", in CAMBOU B. (ed.), *Behaviour of Granular Materials*, Springer, Vienna, 1998.

[VOG 97] VOGEL H.J., "Morphological determination of pore connectivity as a function of pore size using serial sections", *European Journal of Soil Science*, vol. 48, no. 3, pp. 365–377, 1997.

[VUQ 00] VU-QUOC L., ZHANG X., WALTON O.R., "A 3-D discrete-element method for dry granular flows of ellipsoidal particles", *Computer Methods in Applied Mechanics and Engineering*, vol. 187, no. 3, pp. 483–528, 2000.

[VUQ 04] Vu-Quoc L., Lesburg L., Zhang X., "An accurate tangential force–displacement model for granular-flow simulations: contacting spheres with plastic deformation, force-driven formulation", *Journal of Computational Physics*, vol. 196, no. 1, pp. 298–326, 2004.

[WAD 35] Wadell H., "Volume, shape, and roundness of quartz particles", *Journal of Geology*, vol. 43, no. 3, pp. 250–280, 1935.

[WAK 58] Wakabayashi T., "Photoelastic method for determination of stress in powdered mass", *Proceedings of the Eighth Japan National Congress for Applied Mechanics*, Japan National Committee for Theoretical and Applied Mechanics, Science Council of Japan, pp. 235–242, 1958.

[WAL 11] Walker D.M., Tordesillas A., Thornton C. *et al.*, "Percolating contact subnetworks on the edge of isostaticity", *Granular Matter*, vol. 13, no. 3, pp. 233–240, 2011.

[WAL 93] Walton O.R., Braun R.L., "Simulation of rotary-drum and repose tests for frictional spheres and sphere clusters", *Proceedings of the Joint DOE/NSF Workshop on Flow of Particulates and Fluids*, Ithaca, pp. 1–18, 1993.

[WAN 07] Wang J., Gutierrez M.S., Dove J.E., "Numerical studies of shear banding in interface shear tests using a new strain calculation method", *International Journal for Numerical and Analytical Methods in Geomechanics*, vol. 31, no. 12, pp. 1349–1366, 2007.

[WAN 01] Wang Q., Lade P.V., "Shear banding in true triaxial tests and its effect on failure in sand", *Journal of Engineering Mechanics*, vol. 127, no. 8, pp. 754–761, 2001.

[WIK 16] Wikipedia Contributors, "Discrete element method," Wikipedia, The Free Encyclopedia, https://en.wikipedia.org/w/index.php?title=Discrete_element_method&oldid=743921872, accessed October 2016).

[WIL 97] Williams J.R., Nabha R., "Coherent vortex structures in deforming granular materials", *Mechanics of Cohesive-Frictional Materials*, vol. 2, no. 3, pp. 223–236, 1997.

[YIM 02] Yimsiri S., Soga K., "Application of micromechanics model to study anisotropy of soils at small strains", *Soils and Foundations*, vol. 42, no. 5, pp. 15–26, 2002.

[YOS 97] Yoshida T., Tatsuoka F., Siddiquee M.S.A. *et al.*, "Shear banding in sands observed in plane strain compression", in Chambon R., Desrues J., Vardoulakis I. (eds.), *Localisation and Bifurcation Theory for Soils and Rocks*, A.A. Balkema, Rotterdam, 1997.

[YOU 73] Youd T.L., "Factors controlling maximum and minimum densities of sands", in Selig E.T., Ladd R.S. (eds.), *Evaluation of Relative Density and Its Role in Geotechnical Projects Involving Cohesionless Soils*, ASTM Special Technical Publication 523, pp. 98–112, 1973.

[ZHA 01] Zhang H.-W., Zhong W.-X., Gu Y.-X., "A new method for solution of 3D elastic-plastic frictional contact problems", *Applied Mathematics and Mechanics*, vol. 22, no. 7, pp. 756–765, 2001.

[ZHA 10] Zhang J., Majmudar T., Tordesillas A. *et al.*, "Statistical properties of a 2D granular material subjected to cyclic shear", *Granular Matter*, vol. 12, no. 2, pp. 159–172, 2010.

[ZHA 13a] ZHANG W., WANG J., JIANG M., "DEM-aided discovery of the relationship between energy dissipation and shear band formation considering the effects of particle rolling resistance", *Journal of Geotechnical and Geoenvironmental Engineering*, vol. 139, no. 9, pp. 1512–1527, 2013.

[ZHA 11] ZHAO J., GUO N., "Signature of anisotropy in liquefiable sand under undrained shear", in BONELLI S., DASCALU C., NICOT F. (eds.), *Advances in Bifurcation and Degradation in Geomaterials*, vol. 11 of *Springer Series in Geomechanics and Geoengineering*, Springer, 2011.

[ZHA 13b] ZHAO J., GUO N., "A new definition on critical state of granular media accounting for fabric anisotropy", *AIP Conference Proceedings*, vol. 1542, pp. 229–232, 2013.

[ZHO 13] ZHOU B., HUANG R., WANG H. *et al.*, "DEM investigation of particle anti-rotation effects on the micromechanical response of granular materials", *Granular Matter*, vol. 15, no. 3, pp. 315–326, 2013.

[ZHU 95] ZHUANG X., DIDWANIA A.K., GODDARD J.D., "Simulation of the quasi-static mechanics and scalar transport properties of ideal granular assemblages", *Journal of Computational Physics*, vol. 121, no. 2, pp. 331–346, 1995.

Index

Printed in the United States
By Bookmasters